CHEMISTRY IN FOCUS

SKILLS AND ASSESSMENT
WORKBOOK

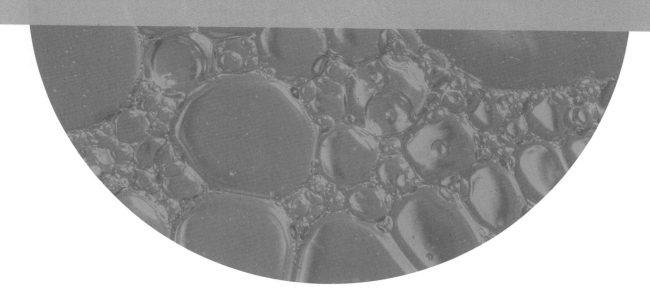

YEAR 11

Debra Smith

Jaya Chowdhury

Samantha Dreon

Chemistry in Focus: Skills and Assessment Workbook Year 11
1st Edition
Debra Smith
Jaya Chowdhury
Samantha Dreon

Publisher: Sam Bonwick
Project editor: Kathryn Coulehan
Editor: Marcia Bascombe
Proofreader: Kelly Robinson
Cover design: Original cover design by Chris Starr (MakeWork), Adapted by
    Justin Lim
Text design: Ruth Comey (Flint Design)
Project designer: Justin Lim
Permissions researcher: Corrina Gilbert
Production controller: Alice Kane
Typeset by: MPS Limited

Any URLs contained in this publication were checked for currency during the
production process. Note, however, that the publisher cannot vouch for the
ongoing currency of URLs.

Acknowledgements
Cover image: iStock.com/ivstiv
Inquiry questions on pages 25, 30, 39, 43, 65, 68, 80, 85, 100, 108, 127, 144,
155 and 165 are from the Chemistry Stage 6 Syllabus © NSW Education
Standards Authority for and on behalf of the Crown in right of the State of
New South Wales, 2017.

For product information and technology assistance,
    in Australia call 1300 790 853;
    in New Zealand call 0800 449 725

For permission to use material from this text or product, please email
aust.permissions@cengage.com

ISBN 978 0 17 044956 4

Cengage Learning Australia
Level 7, 80 Dorcas Street
South Melbourne, Victoria Australia 3205

Cengage Learning New Zealand
Unit 4B Rosedale Office Park
331 Rosedale Road, Albany, North Shore 0632, NZ

For learning solutions, visit cengage.com.au

Printed in China by 1010 Printing International Limited.
1 2 3 4 5 6 7 24 23 22 21 20

CONTENTS

ABOUT THIS BOOK

FEATURES

- Introductory worksheets in Year 11 provide opportunities for you to learn good practices in working scientifically and prepare you to complete high-quality depth studies.

- Review prior knowledge from Stage 5 at the start of each module and check your understanding of key concepts at the end of each module.

- Learning goals are stated at the top of each worksheet to set the intention and help you understand what's required.

- Chapters clearly follow the sequence of the syllabus and are organised by inquiry question.

- Page references to the content-rich student books provide an integrated learning experience.

- Brief content summaries are provided where applicable.

- Hint boxes provide guidance on how to answer questions effectively.

- Fully worked solutions appear at the back of the book to allow you to work independently and check your progress.

ORGANISATION OF YOUR WORKBOOK

An introduction to working scientifically and depth studies

Chapter 1 focuses on the working scientifically skills. These skills are essential for success in the course and we walk you through each one to develop your understanding before applying them to the module content. You may also refer back to this chapter throughout the course.

Chapters 2–15

Each chapter begins with the relevant inquiry question and follows the sequence of the syllabus. Worksheets have been designed to complement the student book and provide additional opportunities to apply and revise your learning. Completion of these worksheets will provide you with a solid foundation to complete assessments and depth studies.

An introduction to working scientifically and depth studies

 Proposing a research question or hypothesis

STUDENT BOOK
Pages 393–5

LEARNING GOALS

Create appropriate research questions and hypotheses.

Identify independent and dependent variables.

Identify variables that need to be controlled.

A good research question identifies the independent and dependent variable in an investigation.
The independent variable is the one that is changed by the scientist while the dependent variable is the variable being investigated and is dependent on the independent variable. For example:

Independent variable   Dependent variable

How does temperature *affect the* rate of a reaction *between 5 mL of 1 mol L^{-1} hydrochloric acid and a 1 cm strip of magnesium ribbon?*

It is important to change only the independent variable and ensure the other variables are controlled.

A hypothesis is a tentative prediction of what is expected based on an existing model or theory. A hypothesis can be tested. It is often written as an 'if … then …' statement. For example:

If the temperature of the 5 mL of 1 mol L^{-1} hydrochloric acid is increased, then the rate of reaction with the 1 cm strip of magnesium ribbon will increase.

1 A teacher provided a Year 11 Chemistry class with some marble chips and 4 mol L^{-1} nitric acid. She asked the class to investigate the relationship between concentration and rate of reaction. The teacher suggested measuring the time taken for the same volume of gas to be produced.

  a Write a research question for the investigation.

_____

_____

  b Write a hypothesis for the investigation.

_____

_____

  c What is the independent variable in the investigation?

_____

  d What is the dependent variable in the investigation?

_____

_____

e Which variables should be controlled?

_____

_____

_____

_____

2 A student performed the same experiment with sodium thiosulfate and hydrochloric acid at three different temperatures. First, she placed 10 mL of $1 \, mol \, L^{-1}$ sodium thiosulfate into each of three separate 50 mL beakers. She filled each of three different test tubes with 5 mL of $2 \, mol \, L^{-1}$ hydrochloric acid and placed one test tube in each of the beakers so it stayed upright. She placed each of the three beaker–test tube systems in water baths at different temperatures for three minutes. She measured and recorded the temperature of the sodium thiosulfate solution for each beaker.

She then did the following with each beaker in turn. She removed the beaker–test tube system from the water bath and placed the beaker on top of a black cross drawn on a white tile. She removed the test tube from the beaker and quickly poured the acid from the test tube into to the beaker. She used a stopwatch to measure the time taken for the cross to disappear (i.e. for the solution to become opaque), as shown below.

a Write a research question for the experiment.

_____

b Write a hypothesis for the experiment.

_____

_____

c What is the independent variable in this investigation?

_____

d What is the dependent variable in this investigation?

_____

e What variables should be controlled?

_____

_____

_____

_____

3 A student was given two white solids labelled A and B. The teacher told the student that both solids were soluble in water. The student's task was to design an investigation to classify the solid as covalent molecular or ionic.

a Write a research question for the experiment.

_____

b Write a hypothesis for the experiment.

_____

4 A class was provided with a table of flame colours for various metal elements. Students were provided with a sample of a solid labelled X and were asked to identify the metal ion that was present in the sample.

Element	Flame colour
Lithium	Carmine (dull red)
Sodium	Yellow
Potassium	Light purple (lilac)
Calcium	Brick-red (orange-red)
Strontium	Scarlet (deep red)
Barium	Pale green (apple green)
Copper	Blue-green

a Write a research question for the experiment.

_____

_____

b Write a hypothesis for the experiment.

_____

_____

5 A student set up two experiments as shown. In experiment 1, he used 100 mL of distilled water. In experiment 2, he dissolved 10 g of salt (NaCl) in 100 mL of distilled water. He recorded the temperature at which the water boiled in each experiment.

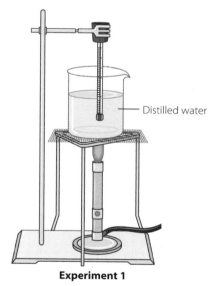

— Distilled water

Experiment 1

a Write a research question for the experiment.

_____

_____

b Write a hypothesis for the experiment.

_____

_____

_____

c What is the independent variable in the investigation?

_____

d What is the dependent variable in the investigation?

_____

e What variables should be controlled?

_____

_____

_____

_____

_____

WS 1.2 Assessing accuracy, precision, reliability and validity in investigations

STUDENT BOOK
Page 396

LEARNING GOALS

Distinguish between types of errors.

Distinguish between accuracy and precision.

Distinguish between reliability and validity.

Calculate the accuracy of data as a percentage.

Label a graph of experimental data.

Accuracy, precision, reliability and validity must be thoroughly understood to avoid errors when processing quantitative data produced from experiments. Some types of errors and uncertainty associated with specific equipment involved in the collection of raw data can be minimised – but not all. There are three basic types of errors when analysing data.

Error	Explanation
Outright mistakes	Poor technique of the operator.
Systematic errors	An error in the system. Always biased in the same direction.
Random errors	Can be in any direction. Minimised or identified by multiple measurements.

The accuracy of data is how close the experimental value is to the theoretical value. It is often represented as a percentage difference. The percentage difference can be calculated using the formula shown.

$$\% \text{ difference} = \frac{|\text{experimental value} - \text{theoretical value}|}{\text{theoretical value}} \times 100$$

1 Identify the type of error in each case.

a	A student read the volume in a measuring cylinder from above instead at eye level.	
b	The top-loading balance was not zeroed before measuring and recording multiple masses.	
c	A student recorded the titre volumes in an experiment as 24.11 mL, 24.15 mL, 24.14 mL and 36.28 mL. What type of error may have contributed to the value of 36.28 mL?	
d	Repetition of an experiment does not minimise this type of error.	
e	Repetition of an experiment does minimise this type of error.	

2 Three groups were given the same mass of an identical substance and asked to record the mass of the substance on a top-loading balance. The actual mass of the substance was 122 g. The results are shown below.

	Trial 1 (g)	Trial 2 (g)	Trial 3 (g)	Trial 4 (g)
Group 1	121	550	120	122
Group 2	132	132	132	132
Group 3	122	122	122	122

a Which group most probably made a mistake? Explain.

_____

_____

b Which group had precise but inaccurate measurements? Explain.

_____

_____

c Which group had precise and accurate measurements?

_____

_____

d Which group may have had a systematic error? Explain.

_____

_____

e Which group had reliable and valid data?

_____

_____

3 A student measures a volume to be 2.0 mL. The actual volume is 2.5 mL. What is the percentage difference in the measurement?

_____

4 Match the diagrams labelled W, X, Y and Z to the statements.

| W | X | Y | Z |

a The aim is precise but not accurate. _____

b The aim is neither precise nor accurate. _____

c The aim is both precise and accurate. _____

d The aim is accurate but not precise. _____

5 A student was investigating the effect on the conductivity of water when various amounts of salt was added to it. Her hypothesis was:

If the amount of salt added to water is increased, then the conductivity will increase.

a Explain whether accuracy or precision is more important for the conductivity meter used to measure conductivity in this experiment.

_____

_____

b The graph of her results is shown by line 2, while line 1 shows the conductance obtained from literature values. The unit for conductance is siemens per metre (S/m), while the mass of salt was measured in grams.

Write a suitable title and label the x and y axes.

HINT

Remember, the independent variable is on the x axis.

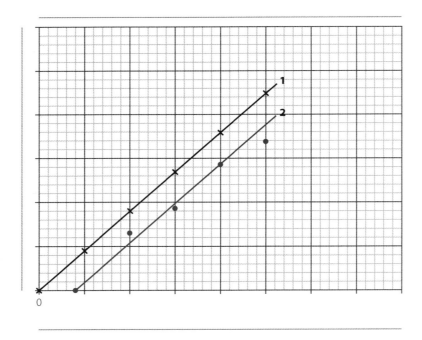

c Assess the reliability, accuracy and validity of the experiment.

_____

_____

_____

_____

d Explain what type of error may have contributed to the experimental results.

_____

_____

WS 1.3 Estimating uncertainties and significant figures

STUDENT BOOK
Pages 399–401

LEARNING GOALS

Determine the uncertainty of measurement when using equipment.

Distinguish between analogue and digital equipment.

Determine the appropriate number of significant figures in data and calculations.

Distinguish between accuracy, precision and reliability in data.

Determine uncertainty in calculations.

The uncertainty range of equipment is stated using the symbol ± and is often found on the equipment. If the uncertainty is not stated on equipment, for analogue measuring equipment, the limit of reading is also sometimes referred to as resolution and it is half the smallest division; for digital measuring equipment it is the smallest scale division.

Uncertainty when calculating average or mean values

There are a number of ways of calculating the average or mean value of data.

Consider the data in the table.

Trial	Mass (g)
1	11.10
2	11.10
3	11.20
4	11.20
Average	11.15

The average value is 0.05 less than the maximum value and 0.05 more than the minimum value. The uncertainty may be quoted as ± 0.05 g.

Therefore, the average may be written as 11.15 g ± 0.05 g.

In reality, when the data set contains a large amount of data, it is unlikely that the above process would be viable. A general rule to calculate the uncertainty of the average is to take two-thirds of the deviation from the average. So, for the above data, two thirds of 0.05 g is 0.03 g; therefore, the average with the uncertainty stated would be 11.15 g ± 0.03 g.

1 a What is the uncertainty stated on the 10 mL graduated pipette?

10 mL x 0.10 B

EX 20°C   tol. ± 0.10

_____

b If a student records the volume of liquid in the graduated pipette as 8.85 mL ± 0.10 mL, what is the range of the true value?

_____

2 a What type of equipment is the voltmeter shown below; i.e. digital or analogue?

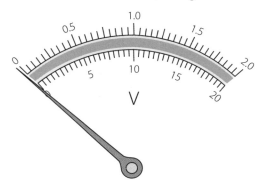

_____

b State the uncertainty when reading the:

   i 2.0 V scale

_____

   ii 20 V scale.

_____

3 a What type of equipment is the top-loading balance shown below; i.e. digital or analogue?

_____

b State the uncertainty of the top-loading balance.

_____

4 The mass of a brass mass labelled 200 g was recorded using four different balances as shown.

          A   **B**   **C**   **D**

a Complete the table for the mass with uncertainty for each balance.

Balance	Uncertainty (±)	Mass (g)	Mass range (g)
A			
B			
C			
D			

b Suggest a reason for the difference in the mass recorded on the various balances.

_____

c When conducting an experiment and recording mass, what does the data suggest about using a balance?

_____

d Which balance is the most precise? Explain your answer.

_____

5 Calculate the change in temperature with uncertainty using the data in the table.

	Temperature (°C) ± 0.01°C
Initial	24.15
Final	63.73

_____

_____

_____

6 A group of three students used a top-loading balance to measure the mass of a beaker. The three students, Alex, Ali and Kim, recorded the mass of the beaker respectively as 39.35 g, 39.36 g and 39.34 g.

The actual mass of the beaker was 38.15 g.

a Record the mass of the beakers in a table showing the absolute uncertainty.

b Comment on the accuracy, precision and reliability of the students' results. Suggest reasons why the three students obtained such different results from the actual value.

_____

_____

_____

_____

_____

_____

_____

_____

_____

_____

_____

c   Calculate the average mass of the beaker using the students' data, including absolute uncertainty in your final answer.

_____

_____

_____

_____

_____

_____

_____

Significant figures

The number of significant figures in data is determined by a set of rules. The basic rules are listed below.

- All non-zero digits count as significant figures. For example, the number 241 has three significant figures.
- Zeroes are significant when they are captive – for example, 241.06 has 5 significant figures – or when they are trailing – for example, 0.500 has 3 significant figures. If there is no decimal point, such as 250, the number of significant figures is not clear because it can be 2 or 3 significant figures. It is, therefore, preferable to write data using scientific notation; for example, such as 2.50×10^2 makes it clear that there are 3 significant figures.

It is important to note that leading zeroes are not significant; for example, 0.06 has only 1 significant figure. When working with data, certain rules apply when using significant figures.

- The final answer cannot have more significant figures than the least number in the question.
- When multiplying or dividing, the final answer should have the same number of significant figures as the factor with the least number of significant figures. For example, $2.32 \times 1.8 = 4.176$. The limiting factor is 1.8; therefore, the final answer should be to two significant figures: 4.2.
- When adding or subtracting, the final answer is dependent on the number with the least decimal places rather than significant figures; for example, $24.1 + 24.16 + 24.167 = 72.427$. The limiting data is 24.1, which has 1 decimal place; therefore, the final answer should be to 1 decimal place: 72.4.

7   Identify the number of significant figures in the numbers given.

a   1.23 _____

b   1.03 _____

c   0.04 _____

d   0.065 _____

e   150 _____

f 2.50×10^2 _____

8   Process the data given below and record your answer to the correct number of significant figures.

a 0.135×202.12

_____

_____

b $135.62 + 51.1 + 42.367$

_____

_____

WS 1.4 Working with data

LEARNING GOALS

Arrange raw data in a table and graphical format.

Calculate an average, median and mode from raw data.

Assess the limitation of using an average value.

Raw data collected during investigations need to be arranged using a range of formats and processed where appropriate. Data can be recorded in tables or represented using appropriate graphs. It is easier to observe trends when using graphs to represent data. Tables and graphs should both have a heading. The independent variable should be in the left-hand column of a table while it should be the 'x' axis on the graph. Columns in the table should have headings and include the units of measurement with uncertainty. Axes on graphs should be labelled and include appropriate units. Data needs to be processed in terms of not only average value but also mode (the value appearing most frequently) and the median (the middle value).

1 A student recorded the results of her experiment in paragraph form as shown below.

Hydrochloric acid was reacted with magnesium and the amount of hydrogen gas produced per minute was 0mL at 0 min, then for every minute after that 0.5mL, 1.0mL, 1.6mL, 2.3mL, 3.0mL, 3.5mL and 4.0mL. The uncertainty in the measurement was ±0.01mL.

a Arrange the data in a table.

b Draw an appropriate graph to represent the data.

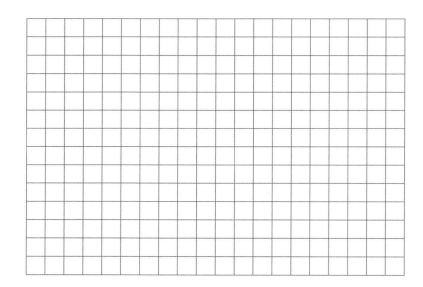

2 A student living on a planet with unusual extreme temperature fluctuations recorded the temperature in her school laboratory over 11 school days. Her purpose was to justify the need for air conditioning in the laboratory because her school principal had said if the average temperature is above 30°C, then she would consider her request.

The temperatures recorded over the period by the student are given below.

55°C, 40°C, 5.0°C, 39°C, 5.0°C, 38°C, 4.0°C, 45°C, 55°C, 3.0°C, 0°C

a Arrange the data using an appropriate format.

b Calculate the average temperature.

_____

_____

c Explain the limitation of using average values in data.

_____

_____

d What is the mode in the data?

_____

e What is the median in the data?

_____

3 The pH of various substances was recorded as shown below.

Substance	pH
Tap water	6
Detergent	8
Vinegar	4
Lemon juice	3
Oven cleaner	10
Table salt	7

Draw an appropriate graph to represent the data.

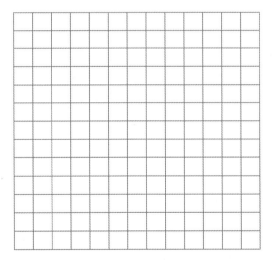

WS 1.5 Performing calculations with your data

STUDENT BOOK
Pages 401–3

LEARNING GOALS

Calculate percentage uncertainties in measurements.

Represent data in a table format and a graphical format.

Analyse data to determine the components of an investigation, including purpose, hypothesis, reliability, accuracy, uncertainty of data and conclusion.

Identify graphs that show scatter plots and to determine whether there is a relationship between the variables.

Absolute uncertainty is stated on equipment.

Percentage uncertainty is expressed as a percentage and is calculated using the formula shown.

$$\text{percentage uncertainty} = \frac{\text{absolute uncertainty}}{\text{measured value}} \times 100$$

When adding or subtracting raw data the absolute uncertainties are added. For example, readings of

$$8.21\,g \pm 0.01\,g \quad + \quad 1.42\,g \pm 0.01\,g \quad = \quad 9.63\,g \pm 0.02\,g$$

When multiplying or dividing, the following steps need to be followed.

▶ Absolute uncertainties need to be first converted to percentage uncertainties for each raw data.

▶ The percentage uncertainties need to be added.

▶ The percentage uncertainty of the calculation needs to be converted to absolute uncertainty.

Tables can be used to organise raw data and processed data, as well as qualitative observations. Tables must have a descriptive title. The independent variable is in the first column and the other columns contain the dependent variable. Each column has a header row with the units and uncertainty stated, where appropriate.

Graphs are drawn to show a pattern in data or a relationship between variables. The independent variable should be on the x axis and the dependent variable should be on the y axis.

1 The four measuring cylinders shown below, 10 mL, 25 mL, 50 mL and 100 mL, were used to measure 10 mL of water. Complete the following table to calculate the percentage uncertainty in each reading.

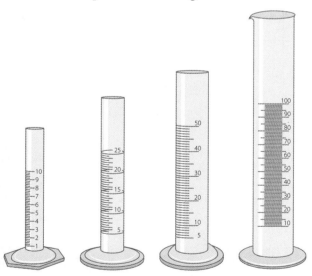

V (measuring cylinder) (mL)	V (measured) (mL)	Absolute uncertainty (mL)	% uncertainty
10	10	±0.2	
25	10	±0.25	
50	10	±0.5	
100	10	±1	

2 Calculate the concentration of a solution in gL^{-1}, when $1.45\,g \pm 0.01\,g$ of NaCl is dissolved in a $250.0\,mL \pm 0.25\,mL$ volumetric flask.

_____

_____

_____

_____

_____

_____

_____

3 A student recorded the following information while reacting magnesium metal with $4\,mol\,L^{-1}$ hydrochloric acid.

Five empty test tubes were labelled V, W, X, Y and Z, placed in five 50 mL beakers and weighed on a top-loading balance. The masses of each test tube and beaker were recorded as 75.079 g, 65.669 g, 62.584 g, 74.745 g and 74.508 g. One magnesium strip was placed in each test tube and the mass was recorded as 75.105 g, 65.695 g, 62.584 g, 74.771 g, and 74.534 g for V, W, X, Y and Z, respectively.

Using a $10\,mL \pm 0.2\,mL$ measuring cylinder, 2 mL, 4 mL, 6 mL, 8 mL and 10 mL of $4\,mol\,L^{-1}$ hydrochloric acid was placed in each test tube. The time taken for the magnesium to disappear was recorded as 29.49 s, 18.72 s, 12.46 s, 7.83 s and 5.01 s, respectively.

a Record the data above in an appropriate table, including the mass of magnesium used in each test tube and uncertainties in the raw data.

b i Suggest a possible research question the student was testing.

_____

_____

ii Suggest a possible hypothesis the student was testing.

_____

_____

c What is the independent variable?

_____

d What is the dependent variable?

_____

e Draw an appropriate graph for the data collected.

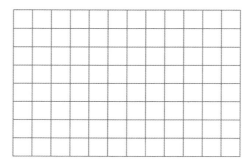

f Explain whether the student's hypothesis was supported.

_____

_____

g Explain whether the data collected was reliable.

_____

_____

h Suggest a suitable conclusion for the experiment.

_____

_____

4 Two students conducted an experiment to determine whether changing the temperature of the hydrochloric acid when it reacts with magnesium affects the time taken for the magnesium to disappear. Each student used the same raw data and plotted their results on the graphs shown below.

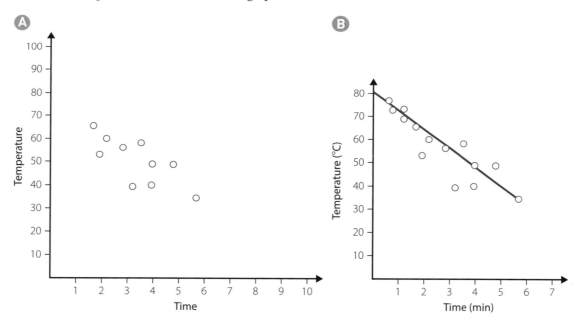

a What is the name given to the types of graphs shown and why are these types of graphs drawn?

_____

b Identify the graph that gives a better representation of the data and explain your answer.

_____

_____

_____

_____

5 There are only three measuring cylinders in a small school laboratory. The volumes and absolute uncertainties of the three measuring cylinders are given in the table below.

V (measuring cylinder)	Absolute uncertainty (±)
10	1
25	2
50	3

Three students with elemental names, Beryllium, Fluorine and Osmium, are given one measuring cylinder each: 10 mL, 25 mL and 50 mL respectively. Their task is to measure out 40 mL of water using their respective measuring cylinders the minimum number of times and pour it into in a 100 mL beaker.

Explain which student will have the least uncertainty in their measured volume of water.

_____

_____

_____

_____

_____

_____

_____

_____

_____

6 A student transferred 15 mL of water using a 50 mL ± 0.50 mL measuring cylinder to a 25 mL ± 0.05 mL measuring cylinder. Assuming there was no loss of water during the transfer, what is the range of the accuracy of the volume in the 25 mL measuring cylinder?

_____

_____

_____

_____

_____

_____

WS 1.6 Interpreting data to solve problems

Problem solving in science involves careful interpretation of both qualitative and quantitative data. The validity and reliability of the data need to be considered when forming justified conclusions.

1 Read the excerpt on the claims for the benefits of a new product, Oxydrate, for hydrating the body and answer the questions that follow.

Oxydrate will change the way you hydrate. Pure liquid oxygen provides more health benefits than drinking pure water. Drink less and hydrate more. Just 30 mL of Oxydrate is equivalent to drinking 600 mL of pure water. No need to carry heavy water bottles anymore while exercising. A study conducted on 10 male athletes aged 18 to 22 revealed that 80% of participants were not thirsty for at least 3.5 hours after drinking Oxydrate. In the study, five athletes were given 30 mL of Oxydrate at room temperature to drink every morning for a week before breakfast while another five athletes were given a placebo, which was 30 mL of pure water. A placebo is a control that contained no Oxydrate. The participants then had to record how long it took before they felt thirsty. The results of the study are given in the table.

Test group	Thirst time (hours)	Placebo group	Thirst time (hours)
A	4.5	F	4.0
B	3.5	G	3.5
C	4.0	H	4.0
D	4.0	I	2.5
E	0.5	J	2.5

a **i** State the claim being made in the excerpt.

_____

ii Explain whether the testing procedure provided evidence for the claim.

_____

_____

b Given that the boiling and melting point of oxygen are −183°C and −219°C respectively, while that of pure water are 100°C and 0°C respectively, evaluate the claim regarding the benefits of drinking 30 mL Oxydrate at room temperature.

_____

_____

_____

_____

c Comment on the reliability and validity of the data.

_____

_____

_____

_____

d Evaluate the statement made in the excerpt '80% of participants were not thirsty for at least 3.5 hours after drinking Oxydrate'.

_____

_____

e Identify all the issues or problems with the excerpt and suggest how you would improve them to test the claim that 30 mL of Oxydrate is better than drinking 600 mL of water.

Issue	Improvement

f A Year 11 student decided to test the contents of Oxydrate. She predicted that, like many sports drinks, Oxydrate contained electrolytes; i.e. ionic substances. When dissolved in water, ionic substances conduct electricity. The data recorded from various tests by her is given below.

Results		
Qualitative	**Quantitative**	
Colourless, odourless liquid	Oxydrate	Boiling point: 105°C
After distillation a white residue remains.		Freezing point: −10°C
When the white residue is heated in a deflagrating		Electrical conductivity: Nil
spoon, it melts and eventually turns black.	White residue after distillation	Melting point: 146°C

i Was the student's prediction correct? Explain your answer.

_____

_____

ii Suggest a possible conclusion the student may make about the contents of Oxydrate.

_____

_____

WS 1.7 Writing using scientific language

LEARNING GOALS

Write a reference using an appropriate format.

Write an appropriate scientific method.

Represent qualitative and quantitative data in a table and/or graphical format.

Interpret information to draw a flow diagram.

Suggest improvements to an experimental procedure.

Communicating in science uses specific scientific language and notation. Experimental procedures are generally recorded in the past passive tense using a sequence of numbered steps in the method. The size and number of equipment, including apparatus and chemicals used, is also recorded. Flow charts are often used to communicate information. In the flow chart, specific shapes are used to convey actions and processes with arrows to show the direction of the flow sequence.

Primary and secondary sources are recorded accurately, with references as appropriate. The style of referencing may be APA, Harvard or one specified by an institution. All forms of reference must include the author or authors' surnames and initials in order of listing in the article, the date published, title of article, place published, name of publisher. If referencing websites, it is important to include the website address, the web page title, date the website was last updated and the date accessed.

1 Write a reference for this workbook.

_____

2 a A student recorded the following method in her practical book.

Method

We used a 100mL measuring cylinder to measure 100mL of water and placed it in a beaker we labelled X. Then we boiled the water in the beaker on a tripod with gauze mat using a Bunsen burner on a blue flame. It started boiling at 99°C. The thermometer we put in the beaker had ±1°C written on it. We then got four more beakers and put 100mL water in them and labelled them A to D. Then we weighed salt on a top-loading balance in four little beakers that we also labelled A to D and put 5.0g of salt, then 10.0g, then 15.0g and finally 20.0g in the small beakers respectively. We then carefully transferred the salt from the little beakers to the water in the corresponding big beakers and dissolved them using a glass rod. My friend set up a tripod and gauze mat above a lit Bunsen burner. We placed a thermometer in each beaker and recorded the temperature at which the salt solution boiled.

Write the method in the past passive tense, including appropriate scientific equipment.

_____

_____

_____

_____

_____

_____

_____

_____

_____

_____

_____

_____

_____

_____

b The student recorded the results as shown below.

Results

The water in all the beakers looked colourless at first and bubbles and steam formed when it started to boil. The water in beaker X boiled at 99°C. The white solid disappeared when we stirred it in beakers A to D.

The solution in beaker A boiled at 103°C, beaker B at 106°C, beaker C at 109°C and beaker D at 115°C.

Write the results in a table, using appropriate scientific language and format.

c Draw an appropriate graph to show how the boiling point of water is related to the mass of salt added.

3 a Read the following passage and construct a flow diagram to summarise the information presented on the scientific method.

The scientific method involves working through a series of steps in a systematic manner.

First, a question is formulated for a certain area of interest; e.g. 'What is the boiling point of pure water at a pressure of 100 kPa?'. Background research is then undertaken to determine what is already known about the topic. The literature reports a boiling point of 99.61°C at this pressure for pure water. The student then formulates the hypothesis: 'If the pressure in the laboratory is 100 kPa, then pure water will boil at 99.61°C'. An investigation is designed to test the boiling point of pure water in the laboratory and the atmospheric pressure is measured and found to be 100 kPa. The experiment is repeated several times and data is gathered and analysed. The average boiling point of water is found to be 101.2°C ± 0.1°C. The student draws conclusions that the evidence does not support the hypothesis. The student then repeats the experiment by amending certain features. This time she finds the average boiling point to be 99.6°C ± 0.1°C. She now submits her laboratory report to her teacher.

b Suggest two possible amendments the student may have made to her experiment to obtain the correct result.

_____

_____

PROPERTIES AND STRUCTURE OF MATTER

Reviewing prior knowledge

1 Match each term on the left with the appropriate description on the right by writing the number of the description next to the term.

A	Homogeneous		1	Is a physical property
B	Heterogeneous		2	Is a chemical property
C	Mixture		3	Is the number of protons
D	Pure substance		4	Is the number of protons plus neutrons
E	Compound		5	Are made up of protons, neutrons and electrons
F	Element		6	Have the same chemical properties
G	Sublimation		7	Atoms of the same element that have different number of neutrons
H	Solution		8	Is the only metal that is a liquid at room temperature
I	Suspension		9	Arises from the decay of nuclei in atoms
J	Boiling point		10	Is a diatomic element
K	Decomposition		11	Is made up of molecules
L	Atomic number		12	Milk is an example of this
M	Mass number		13	Is a homogeneous mixture
N	Atoms		14	Is the term used when a solid changes to a gas
O	Elements in groups in periodic table		15	Is a pure substance that cannot be decomposed into simpler substances
P	Isotope		16	Is a physical combination of substances
Q	Mercury		17	Means non-uniform or variable composition
R	Radioactivity		18	Is a substance that is not contaminated by any other substance
S	Hydrogen gas		19	Means uniform composition throughout
T	Water		20	Is a pure substance that is made up of two or more elements that are chemically combined in fixed ratios

2 Identify the separation technique for each mixture using the words from the list below.

decantation	filtration	sedimentation
distillation	fractional distillation	separating funnel

Mixture	Separation technique
Crude oil	
Oil and water	
River sand with gold	
Salt water	
Sand in water	
Sawdust and water	

3 Calculate the percentage composition of a mixture that is made up of 12.5 g of salt, 11.6 g of sand and 25.9 g of iron filings.

_____

_____

_____

4 Complete the table.

Name	Formula
Oxygen gas	
Aluminium oxide	
Carbon dioxide	
	Co
	NH_3
	Cl_2
	Ag

5 Complete the table.

Formula	How many different elements?	What are the elements?	Total number of atoms	Made up of ions/molecules
$NaHCO_3$				
$Mg(OH)_2$				
Li_3PO_4				
$KMnO_4$				
CCl_4				

2.1 Classifying and separating matter

STUDENT BOOK
Pages 23–30,
33–40

LEARNING GOALS

Identify a mixture as homogenous or heterogenous based on physical properties.

Separate components of a heterogenous mixture based on physical properties.

Classify elements based on their physical properties.

Describe the use of an element based on its physical property.

Identify groups, periods and transition elements in the periodic table.

Apply IUPAC nomenclature to inorganic substances.

Identify the location of specific elements in groups and periods in the periodic table.

1 Four unknown substances were labelled W, X, Y and Z. Their properties are listed in the table below.

   a Classify them as metal, non-metal or semi-metal.

Substance	Appearance	Malleable/brittle	Melting point (°C)	Electrical conductivity (S m^{-1})	Classification
W	Shiny, salmon pink	Malleable	1083	5.9×10^7	
X	Dull, grey black	Brittle	3730	1.0×10^5	
Y	Dull, yellow	Brittle	98	1.0×10^{-15}	
Z	Shiny grey	Brittle	1414	1.0×10^3	

   b Identify the element that could be used in electrical wiring. Give a reason for your answer.

2 S and T are two immiscible liquids with densities of $1\,\text{g}\,\text{mL}^{-1}$ and $0.8\,\text{g}\,\text{mL}^{-1}$ respectively. Draw a labelled diagram to show how the mixture can be separated to obtain S and T.

Matter can be classified based on chemical and physical properties. Separation of mixtures is based on physical properties of the components in the mixture.

3 A student was given a mixture of four substances A, B, C and D. The properties of each substance are listed below.

Substance	State at room temperature	Solubility in			Boiling point (°C)	Density (g mL^{-1})
		water	C	D		
A	Solid	x	x	x	4827	2.6
B	Solid	√	√	x	1465	2.2
C	Liquid	√		x	100	1.0
D	Liquid	x	x		68	0.66

a   Is the mixture homogeneous or heterogenous? Explain your answer.

_____

_____

_____

_____

_____

b   Complete the flow chart to show how to separate the components so that each component can be collected.

Mixture of A, B, C, D

4 a In the periodic table shown, label the groups, periods, the transition elements and the main group elements.

1																	18
1 H	2											13	14	15	16	17	2 He
3 Li	4 Be											5 B	6 C	7 N	8 O	9 F	10 Ne
11 Na	12 Mg	3	4	5	6	7	8	9	10	11	12	13 Al	14 Si	15 P	16 S	17 Cl	18 Ar
19 K	20 Ca	21 Sc	22 Ti	23 V	24 Cr	25 Mn	26 Fe	27 Co	28 Ni	29 Cu	30 Zn	31 Ga	32 Ge	33 As	34 Se	35 Br	36 Kr
37 Rb	38 Sr	39 Y	40 Zr	41 Nb	42 Mo	43 Tc	44 Ru	45 Rh	46 Pd	47 Ag	48 Cd	49 In	50 Sn	51 Sb	52 Te	53 I	54 Xe
55 Cs	56 Ba	57-71	72 Hf	73 Ta	74 W	75 Re	76 Os	77 Ir	78 Pt	79 Au	80 Hg	81 Tl	82 Pb	83 Bi	84 Po	85 At	86 Rn
87 Fr	88 Ra	89-103	104 Rf	105 Db	106 Sg	107 Bh	108 Hs	109 Mt	110 DS	111 Rg	112 Cn	113 Nh	114 Fi	115 Mc	116 Lv	117 Ts	118 Og

58 La	58 Ce	59 Pr	60 Nd	61 Pm	62 Sm	63 Eu	64 Gd	65 Tb	66 Dy	67 Ho	68 Er	69 Tm	70 Yb	71 Lu
89 Ac	90 Th	91 Pa	92 U	93 Np	94 Pu	95 Am	96 Cm	97 Bk	98 Cf	99 Es	100 Fm	101 Md	102 No	103 Lr

b Complete the information below using the periodic table given above.

	Description	Answer
i	Period 2, group 2	
ii	Very reactive metal in period 4	
iii	Most metallic character in group 1	
iv	The period(s) with only one semi-metal	
v	The period(s) with only gases	
vi	The group with alkali metals	
vii	The group with alkaline earth metals	
viii	The group with the halogens	
ix	The group with the noble gases	

5 Complete the table.

Name	Formula
	Cu_2O
	$CuSO_4$
Sodium carbonate	
Magnesium hydroxide	
	Fe_2O_3
	SCl_2
Carbon monoxide	
Dinitrogen pentoxide	

Apply gravimetric analysis to determine mass of components in a mixture.

Calculate the percentage composition of mixtures.

Calculate uncertainty in measurement.

1 a A mixture contained sand, salt and iron filings. Describe a numbered procedure for performing gravimetric analysis on this mixture.

_____

_____

_____

_____

_____

_____

_____

_____

_____

_____

_____

_____

b A student recorded the following data from the gravimetric analysis.

Component	Mass (g) ±0.1 g	Component	Mass (g) ±0.1 g
Empty 100 mL beaker	42.1	Empty 250 mL conical flask	162.1
100 mL beaker + mixture	99.4	Conical flask + filtrate	261.9
Magnet	2.1	Filter paper + sand	49.5
Magnet + iron filings	8.0	Empty evaporating basin	123.7
Filter paper	2.2	Evaporating basin + salt	127.8

i Calculate the mass of the mixture and each component and state the uncertainty of each mass.

Mass(mixture): _____

Mass(iron filings): _____

Mass(sand): _____

Mass(salt): _____

ii Calculate the percentage composition of the mixture. Include percentage uncertainties for each mass component in your answer.

_____

_____

_____

_____

_____

_____

_____

_____

_____

_____

_____

2 Iron ores are rocks that usually contain haematite, Fe_2O_3, and magnetite, Fe_3O_4. A one tonne sample of ore was found to contain 100 kg of haematite and 990 g of magnetite. Determine the percentages of haematite and magnetite in the rock. (Note: 1 tonne $= 10^6$ g)

Haematite =   Magnetite =

_____

3 Sinus and nasal decongestants often contain an active ingredient called pseudoephedrine hydrochloride. Its formula is $C_{10}H_{15}NO.HCl$.

a A 5.2 g tablet was found to contain 1.1% of pseudoephedrine hydrochloride. Calculate the mass of pseudoephedrine hydrochloride in the tablet.

_____

b The maximum permissible dose of pseudoephedrine hydrochloride must not exceed 240 mg per 24 hours. Based on your answer in part **a**, what is the maximum number of tablets that can be consumed in 24 hours?

_____

4 Equipment used in public gyms, such as weights and dumbbells, tend to be places for transmission of bacterial infections. A study found that increasing the percentage of copper in equipment reduces the amount of surface bacteria. For economic reasons, using 65% copper alloy reduced the amount of surface bacteria to an acceptable level on the equipment. A company produced 2.0 kg dumbbells and advertised it to be effective in reducing bacteria. The copper content was stated as 570 000 ppm per dumbbell (ppm = parts per million). Would these dumbbells be suitable for reducing bacteria?

_____

_____

_____

_____

_____

3 Atomic structure and atomic mass

 WS 3.1 Identifying atoms of elements from the number of subatomic particles

STUDENT BOOK
Pages 45–53

LEARNING GOALS

Identify elements by their atomic number.

Describe how electrons, protons and neutrons are distributed in atoms.

Interpret information related to isotopes of elements.

Identify and apply factors that determine the stability of isotopes.

1 A student wrote the following paragraph but it contains a number of mistakes. Rewrite the paragraph so it is correct.

An atom is made up of a nucleus that is large and light because it holds protons and electrons. The protons have a negative charge while the electrons have a positive charge. The nucleus has no charge because the number of electrons is the same as the number of protons. The nucleus is surrounded by a cloud of neutrons. The bulk of the volume of the atom is from the nucleus. The nucleus has a diameter about ten thousand times the diameter of the whole atom. An element is made up of lots of different types of atoms. The atomic number, with symbol A, refers to the number of neutrons while the mass number, with symbol Z, refers to the number of protons. The term 'nucleon' is used to describe either a proton or an electron. The atomic number is sometimes called the nucleon number. Isotopes of elements are atoms that have different numbers of electrons in their nuclei. Isotopes of the same element will have the same mass number but different atomic numbers.

_____

_____

_____

_____

_____

_____

_____

_____

_____

_____

_____

_____

_____

2 There are 37 known isotopes of iodine. Complete the missing items in the table for the two isotopes shown.

Isotope	Symbol	Atomic number	Mass number	Number of protons	Number of neutrons	Number of electrons
Iodine-125	I					
	I⁻		131			

3 Complete the paragraphs using the words/numbers given. A word/number may be used more than once or not at all.

1:5	13	equal	protons
1:6	82	less	radiation
1	125	greater	radioisotopes
5	132	high	spontaneously
6	207	increases	stable
7	alpha	not	stability
12	beta	neutron	unstable

Isotopes of elements may be stable or _____. Unstable isotopes tend to _____ emit _____

and are referred to as _____ or as being radioactive isotopes. The radioactive emissions come from

the _____ nucleus. The _____ of the nucleus is determined by the neutron to proton ratio.

If the ratio of neutrons to protons is too _____ or too low, the nucleus may be _____.

Elements with an atomic number _____ than or equal to 20 tend to have approximately _____

numbers of protons and neutrons and the neutron to proton ratio is close to _____ for stable isotopes.

If the neutron to proton ratio is _____ close to 1, then these isotopes are _____. For example,

carbon-12, has _____ neutrons and 6 protons. Its _____ to proton ratio is 6 to 6; therefore, it

is _____. However, carbon-13, has _____ neutrons and _____ protons; therefore,

its neutron to proton ratio is _____ to _____, which is _____ equal to 1, making

it _____.

As the atomic number _____, the neutron to proton ratio for a _____ nucleus

becomes _____ than 1 and is close to 1:5. For elements with atomic number _____ than 20

and _____ than 83, if the neutron to proton ratio is _____ equal to the ratio for the stable

elements in the periodic table or about 1:5, then the nucleus is _____. For example, chromium-52

is stable because it is the isotope stated in the periodic table and has a neutron to proton ratio of about 1:2.

However, chromium-54 is _____ as its neutron to proton ratio is 1:3, which is not 1:2 as listed for

the stable isotope. Another example, lead-207, has _____ neutrons and 82 protons. Its neutron to

proton ratio is close to _____; therefore, it is _____. Lead-214 has _____ neutrons

and _____ protons. Its neutron to proton ratio is close to _____; therefore, it is _____.

All nuclei with an atomic number _____ than 83 are _____. These nuclei contain too many protons and neutrons and this results in their instability. They tend to emit an _____ particle, which reduces the number of _____ and neutrons by two.

4 Complete the table for each isotope shown.

Isotope	Neutron to proton ratio or $Z > 83$, too big	Stable or unstable
$^{3}_{1}H$		
$^{20}_{10}Ne$		
$^{232}_{90}Th$		
$^{37}_{17}Cl$		
$^{231}_{91}Pa$		
$^{197}_{79}Au$		
$^{18}_{8}O$		
$^{238}_{92}U$		

5 In the periodic table given below, shade in the elements that:

a have unstable isotopes if the neutron to proton ratio is not close to 1

b always have unstable isotopes.

1																	18
1 H	2											13	14	15	16	17	2 He
3 Li	4 Be											5 B	6 C	7 N	8 O	9 F	10 Ne
11 Na	12 Mg	3	4	5	6	7	8	9	10	11	12	13 Al	14 Si	15 P	16 S	17 Cl	18 Ar
19 K	20 Ca	21 Sc	22 Ti	23 V	24 Cr	25 Mn	26 Fe	27 Co	28 Ni	29 Cu	30 Zn	31 Ga	32 Ge	33 As	34 Se	35 Br	36 Kr
37 Rb	38 Sr	39 Y	40 Zr	41 Nb	42 Mo	43 Tc	44 Ru	45 Rh	46 Pd	47 Ag	48 Cd	49 In	50 Sn	51 Sb	52 Te	53 I	54 Xe
55 Cs	56 Ba	57-71	72 Hf	73 Ta	74 W	75 Re	76 Os	77 Ir	78 Pt	79 Au	80 Hg	81 Tl	82 Pb	83 Bi	84 Po	85 At	86 Rn
87 Fr	88 Ra	89-103	104 Rf	105 Db	106 Sg	107 Bh	108 Hs	109 Mt	110 DS	111 Rg	112 Cn	113 Nh	114 Fi	115 Mc	116 Lv	117 Ts	118 Og

58 La	58 Ce	59 Pr	60 Nd	61 Pm	62 Sm	63 Eu	64 Gd	65 Tb	66 Dy	67 Ho	68 Er	69 Tm	70 Yb	71 Lu
89 Ac	90 Th	91 Pa	92 U	93 Np	94 Pu	95 Am	96 Cm	97 Bk	98 Cf	99 Es	100 Fm	101 Md	102 No	103 Lr

Describing energy levels for electrons in atoms and ions

LEARNING GOALS

Describe energy levels, orbitals, shells and sublevels.

Describe the number of electrons found in energy levels and orbitals.

Recognise the shapes of *s*, *p* and *d* orbitals.

Apply electron configuration in terms of *spdf* notation.

Distinguish between an absorption and an emission spectrum.

Describe how a hydrogen discharge tube works.

Interpret spectral evidence for elements to identify presence of a specific element in a sample.

1   a   Match the terms from the list to the parts labelled 1 to 11 in the two diagrams below.

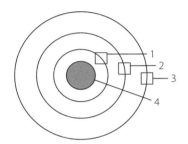

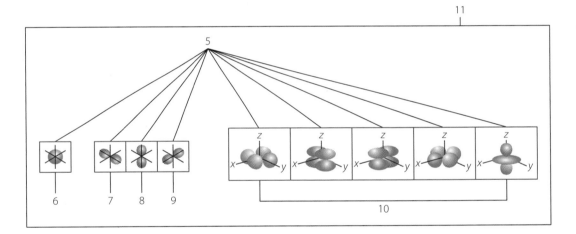

	Term	Label number		Term	Label number
i	*d*		vii	Main energy level	
ii	*s*		viii	Energy level 1	
iii	p_x		ix	Energy level 2	
iv	p_y		x	Energy level 3	
v	p_z		xi	Nucleus	
vi	Orbital				

b State the number of sublevels found in the main energy levels listed below, their identity and the maximum number of electrons each energy level can accommodate.

The first one has been done for you.

Main energy level	Number of sublevels	Identity of sublevels	Maximum number of electrons
1	1	1s	2
2			
3			
4			

c State the maximum number of electrons an orbital can hold.

_____

d The shape of 1s, 2s and 3s orbitals are all spherical. What is the difference between them?

_____

2 a Write the electron configuration for the following using *spdf* notation.

   i nitrogen

_____

   ii lithium ion (Li$^+$)

_____

   iii calcium

_____

   iv chloride ion (Cl$^-$)

_____

b A species has electron configuration $1s^2 2s^2 2p^6$. Suggest its possible identity. (NB: there may be more than one answer.)

_____

_____

3 An atomic absorption spectrum is complementary to an atomic emission spectrum. Both types of spectra, labelled A and B, are shown below. Identify each type of spectrum.

STUDENT BOOK
Pages 64–5

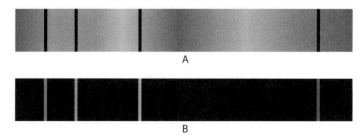

A

B

A = _____

B = _____

4 When a hydrogen gas discharge tube has a high voltage passed through it, the tube starts to glow a bright purple-pink colour. When the glowing tube is viewed through a spectroscope or a prism, the following spectrum is observed.

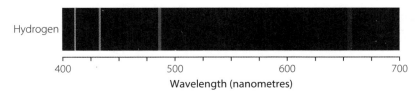

Hydrogen

Wavelength (nanometres)

a What is the purpose of applying the high voltage to the hydrogen discharge tube?

_____

_____

_____

b Explain why the purple-pink colour appears when viewed through a spectroscope or prism.

_____

_____

_____

_____

_____

_____

c What do the lines in the spectrum correspond to?

_____

_____

_____

_____

_____

d Rank the lines in the spectrum above from 1 to 4, where 1 is the biggest jump in energy and 4 the smallest jump.

_____

_____

5 The emission spectra of a mixture of elements is shown as well as the emission spectra of some known elements.

a Identify the elements present in the mixture.

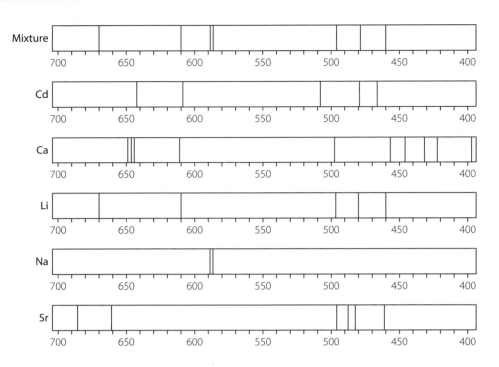

b A flame test was done on the mixture. Explain the colour of flame expected for the mixture.

STUDENT BOOK
Pages 68–9

WS 3.3 Investigating the properties of unstable isotopes

STUDENT BOOK
Pages 50–4

LEARNING GOALS

Identify the penetrating power of alpha, beta and gamma rays.

Apply the concept of half-life to solve problems.

Write balanced nuclear equations.

Analyse the use and effects of radioisotopes.

1 Identify the types of radiation (i, ii and iii) shown in the diagram, based on their penetration power.

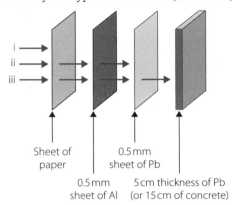

i _____

ii _____

iii _____

2 The half-life of a radioisotope is the time required for half the atoms in a given sample to undergo radioactive decay. A radioactive source in the shape of a disc was placed on a few sheets of newspaper. Under the newspaper was some photographic film. After 10 minutes, an image in the shape of the disc had developed on the film.

Then, after 16 days, when the same disc was placed on a few sheets of newspaper on top of photographic film again, the image on the photographic film took 40 minutes to develop to the same intensity.

Using the table, deduce the identity of the radioactive source. Explain your choice.

Radioisotope	Type of radiation emitted	Half-life	Radioisotope	Type of radiation emitted	Half-life
a	α, β	10 minutes	d	β, γ	8 days
b	β, γ	40 minutes	e	α	16 days
c	α	8 days	f	β, γ	16 days

_____

_____

_____

_____

_____

_____

3 Write balanced nuclear reactions for the following.

a Polonium-214 undergoes alpha decay.

_____

b Carbon-14 undergoes beta decay.

c Uranium-238 is first bombarded with neutrons then it undergoes beta decay.

_____

d Plutonium-239 is bombarded with neutrons to produce plutonium-241, which undergoes decay to produce americium-241.

_____

4 Iodine-123 and iodine-131 are both used in medicine. Iodine-123 is used for diagnosis while iodine-131 is used mainly to treat certain cancers of the thyroid. Patients' exposure to radioisotopes is generally kept at minimal levels to prevent side effects. The properties of each radioisotope are listed in the table, as well as the type of radiation produced during the decay process. NB: Beta radiation is not able to pass through human tissue very well and tends to transfer some of its energy to tissues to kill cells, but gamma radiation can easily pass through the body, with minimal damage to tissue.

Radioisotope	Type of radiation emitted during decay	Half-life
Iodine-123	γ	13 hours
Iodine-131	β, γ	8 days

a Justify the use of each isotope in medicine with respect to the properties listed in the table.

_____

_____

_____

_____

b Write an equation to show the decay of iodine-131.

_____

5 Radium-226 is a naturally occurring radioisotope that results from the natural radioactive decay of uranium. Radium-226 was mixed with paint and other chemicals to paint watch faces until the 1960s, so they would glow a greenish colour, making the watches particularly useful in the dark. It was mainly female factory workers, known as 'Radium Girls', who carried out the job of painting the watch dials. They were advised to make the paint brushes pointy by placing them on their lips. The women often used the radium to paint their nails because they were told it was harmless. The radium emitted all three types of radiation. Alpha radiation is ionising radiation and can potentially harm biological molecules. Beta radiation can kill cells in tissues while gamma radiation can pass straight through the body. Radium has a half-life of 1600 years.

a Write an equation to show the alpha decay of radium-226.

_____

b Suggest one health effect of radium-226 on the 'Radium Girls', giving a reason for your answer.

_____

_____

c The many clocks and watches that were made in the early 20th century are no longer in use. However, they still pose a threat now in the 21st century. Explain why these radium watches still pose a threat and how they may have been disposed of in the 1960s.

_____

_____

_____

_____

4 Periodicity

WS 4.1 Analysing patterns in properties of elements

STUDENT BOOK
Pages 75–84

LEARNING GOALS

Identify the states of the elements in the periodic table at 25°C and 100 kPa.

Explain and predict the trends in the physical and chemical properties of elements in the periodic table, including atomic radius, ionisation energy, electronegativity and reaction with water.

Define ionisation energy and write equations for successive ionisations.

Identify the energy level and the orbital an electron is being removed from during ionisation of an atom.

1 Using the key provided, indicate the state of the elements at 25°C and 100 kPa on the periodic table.

Solid	◹
Liquid	●
Gas	▨

1																	18
1 H	2											13	14	15	16	17	2 He
3 Li	4 Be											5 B	6 C	7 N	8 O	9 F	10 Ne
11 Na	12 Mg	3	4	5	6	7	8	9	10	11	12	13 Al	14 Si	15 P	16 S	17 Cl	18 Ar
19 K	20 Ca	21 Sc	22 Ti	23 V	24 Cr	25 Mn	26 Fe	27 Co	28 Ni	29 Cu	30 Zn	31 Ga	32 Ge	33 As	34 Se	35 Br	36 Kr
37 Rb	38 Sr	39 Y	40 Zr	41 Nb	42 Mo	43 Tc	44 Ru	45 Rh	46 Pd	47 Ag	48 Cd	49 In	50 Sn	51 Sb	52 Te	53 I	54 Xe
55 Cs	56 Ba	57-71	72 Hf	73 Ta	74 W	75 Re	76 Os	77 Ir	78 Pt	79 Au	80 Hg	81 Tl	82 Pb	83 Bi	84 Po	85 At	86 Rn
87 Fr	88 Ra	89-103	104 Rf	105 Db	106 Sg	107 Bh	108 Hs	109 Mt	110 DS	111 Rg	112 Cn	113 Nh	114 Fi	115 Mc	116 Lv	117 Ts	118 Og

58 La	58 Ce	59 Pr	60 Nd	61 Pm	62 Sm	63 Eu	64 Gd	65 Tb	66 Dy	67 Ho	68 Er	69 Tm	70 Yb	71 Lu
89 Ac	90 Th	91 Pa	92 U	93 Np	94 Pu	95 Am	96 Cm	97 Bk	98 Cf	99 Es	100 Fm	101 Md	102 No	103 Lr

2 a Indicate by using arrows the trend in atomic radius across a period and down a group in the periodic table.

b Explain the reasons for the trends you identified in part **a**.

_____

_____

_____

_____

_____

_____

_____

_____

_____

_____

_____

_____

_____

_____

_____

_____

c Compare the radius of the sodium ion, Na^+, with that of the oxide ion, O^{2-}, and explain your response.

_____

_____

_____

_____

3 a Define the term 'ionisation energy' and state its units.

_____

_____

b Draw a sketch to show the relative successive ionisation energies of boron.

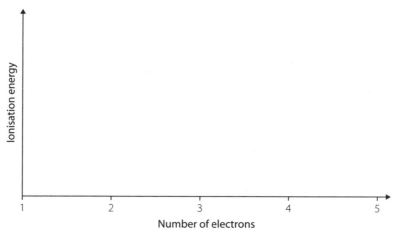

c Write equations to show the successive ionisation of boron shown in your graph in part **b** above.

1st ionisation: _____

2nd ionisation: _____

3rd ionisation: _____

4th ionisation: _____

5th ionisation: _____

d **i** State the electron configuration for a neutral boron atom using *spdf* notation.

_____

ii For each ionisation in part **c** above, identify the energy level and the orbital the electron is being removed from.

1st ionisation: _____

2nd ionisation: _____

3rd ionisation: _____

4th ionisation: _____

5th ionisation: _____

4 a Define the term 'electronegativity'.

_____

_____

_____

b Describe the trend in electronegativity in the periodic table, across a period and down a group.

_____

_____

c Noble gases are assigned an electronegativity value of zero, while fluorine is assigned the maximum value for electronegativity of 3.98. What is the name of this scale that is used to assign the values?

_____

5 Insert the appropriate terms at the end of each statement.

a All elements in this group react with cold water. _____

b Only some elements in this group react with cold water. _____

c This element reacts with steam at high temperatures. _____

d Elements that react with water to produce a basic solution are metals. True or False? _____

e The compound that is formed when sodium reacts with water. _____

f The element that is formed when sodium reacts with water. _____

g Elements that react with water are located on the right-hand side of the periodic table. True or False? _____

h The smaller the atomic radius, the greater the reactivity with water. True or False? _____

6 The table summarises the tests for four gases and their reactions with water. Moistened litmus paper turns blue in basic and red in acidic solutions. Refer to the table to answer the questions.

Element	Test for gas	Reaction with water
Chlorine	Turns moistened litmus paper red	$Cl_2(g) + H_2O(l) \rightarrow HOCl(aq) + HCl(aq)$
Fluorine	Turns moistened litmus paper red	$F_2(g) + H_2O(l) \rightarrow O_2(g) + HF(aq)$
Hydrogen gas	A lit splint goes 'pop'	No visible reaction
Oxygen gas	Re-ignites a glowing splint	No visible reaction

a Which elements in the table are soluble in water? Explain your response.

_____

b Jade performed an experiment to determine the identity of a gas labelled X. Her teacher told her it was a gas from the table given above. Jade bubbled the unknown gas, labelled X, through neutral water. She collected the gas in an inverted tube as shown. When the collected gas was tested with a glowing splint, it re-ignited the splint.

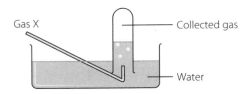

Jade concluded that gas X must be oxygen. Evaluate the validity of her conclusion and suggest improvements to her procedure.

_____

_____

_____

_____

_____

_____

_____

_____

_____

_____

_____

_____

_____

_____

_____

Bonding

STUDENT BOOK
Pages 90–8
101–3

WS 5.1 Determining types of bonds

LEARNING GOALS

Identify the type of bond formed by the difference in electronegativity of elements.

Identify how ions form.

Write Lewis structures for formation of ionic bonds.

Determine the valency of elements in the periodic table.

Write formulae and names of ionic and covalent compounds.

Atoms have a tendency to achieve stable electron configurations. They do this by bonding with other atoms. Bonding also lowers the potential energy between the particles, which can be oppositely charged ions or nuclei and the electrons between them. The difference in electronegativity, ΔEN, of the atoms in a bond will result in an uneven sharing of their electrons. The difference in electronegativity can be used to determine the type of bond formed. The following rules can be applied to determine the type of bond formed:

$\Delta EN = 0$ Pure covalent

$0 \leq \Delta EN \leq 1.8\text{–}2.0$ Polar covalent

$\Delta EN \geq 1.8\text{–}2.0$ Ionic

				H 2.20				He 0
Li 0.98	Be 1.57		B 2.04	C 2.55	N 3.04	O 3.44	F 3.98	Ne 0
Na 0.93	Mg 1.31		Al 1.61	Si 1.90	P 2.19	S 2.58	Cl 3.16	Ar 0
K 0.82	Ca 1.00			As 2.18	Se 2.55	Br 2.96	Kr 0	
Rb 0.82	Sr 0.95					I 2.66	Xe 0	
Cs 0.79	Ba 0.89						Rn 0	

1 Determine the type of bond formed in each of the substances listed – i.e. pure covalent, polar covalent or ionic – with reference to the data provided above.

a CsCl: _____

b CaO: _____

c CO: _____

d Cl_2: _____

e HCl: _____

2 Ions form by the exchange of electron(s) between a metal and a non-metal atom.

   a   **i**   Write a balanced equation to show the reaction between sodium metal and chlorine gas to form sodium chloride solid.

          _____

      ii   Use Lewis structures to write the above equation.

          _____

   b   **i**   Write a balanced equation to show the reaction between magnesium metal and oxygen gas to form magnesium oxide solid.

          _____

      ii   Use Lewis structures to write the above equation.

          _____

3 **a** Valence or valency of an element is a number that measures the combining power of the element when it forms compounds. Valency is usually a positive number, although the term 'valence' can sometimes refer to the actual charge on the ionic species.

      Determine the valency of the species listed.

Species	Valency
Ag	
C	
F	
He	
N	
O	
Zn	

   b Some metals display variable valencies as they can form more than one type of ion. These valencies are indicated as capital Roman numerals in brackets. Complete the table to show the variable valencies of the metal elements listed.

Metal element	Ions formed	Valency	Name with valency
Copper			
Gold			
Iron			
Lead			
Mercury			
Tin			

c Complete the table below.

Name	Formula
Copper(II) sulfate	
Gold(III) chloride	
Iron(II) hydroxide	
Mercury(II) oxide	
	PbI_4
	SnS
	Hg_2Br_2
	Cu_2O

4 In covalent compounds, the valency or valence of an element is the number of covalent bonds the element forms. Non-metal elements can also display multiple valencies.

a For each covalent compound listed, identify the valency of the underlined element.

Compound	Valency of underlined element
H$\underline{F}$	
$\underline{N}H_3$	
$H_2\underline{S}$	
$\underline{C}O_2$	
$\underline{P}Cl_5$	

b Complete the table below.

Name	Formula
	N_2O_5
	SF_6
Carbon tetrachloride	
Sulfur dioxide	
Sulfur tetrafluoride	SF_4
Nitrogen trichloride	NCl_3
Ammonia	
Silicon dioxide	

STUDENT BOOK
Page 104–11

Identify the chemical structures of ionic, covalent network, covalent molecular and metallic substances.

Identify physical properties of ionic, covalent network, covalent molecular and metallic substances.

1 Use the appropriate numbers from the table below to complete the flow diagram on the following page. A term may be used more than once. To keep it simple, write the numbers from the table in the diagram.

Number	Term	Number	Term
1	Have very high melting and boiling points	15	e.g. SiO_2
2	Covalent bonding extends indefinitely through the lattice	16	e.g. H_2O
3	Made up of ions of opposite charges	17	e.g. Mg
4	Electrostatic forces of attraction between oppositely charged ions	18	e.g. NaCl
5	Held in fixed lattice positions	19	Have low melting and boiling points
6	Exists as discrete molecules	20	Are malleable and ductile
7	Made up of three-dimensional array of positive ions held together in a sea of delocalised electrons	21	Are hard and brittle
8	Conduct electricity when molten or dissolved in water	22	Electrons carry the electric charge
9	Conduct electricity in the solid state	23	Ions carry the electric charge
10	Do not conduct electricity	24	Mostly solid at room temperature with one exception
11		25	Can be solid, liquid or gas at room temperature
12		26	Solid at room temperature
13		27	Have high melting and boiling points
14			

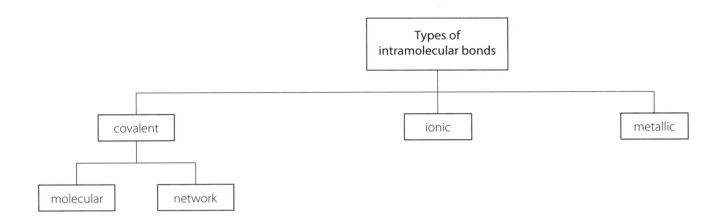

2 Choose from the list below to complete the table.

| Ag | brittle | electrons | KCl | permanent dipoles |
| atoms | dull | ions | lustrous | temporary dipoles |

Description	Term
Sulfur appears …	
Copper appears …	
Carry the charge in graphite	
Carry the charge in molten NaCl	
Hydrogen molecule is made up of …	
Methane molecule has …	
Hydrogen sulfide has …	
Is made up of ions	
Can conduct electricity in the solid form	
Magnesium oxide is …	

Determine shapes of molecules using valence shell electron-pair repulsion theory.

Define allotropes.

Identify allotropes given their shape and their properties.

Identify uses of carbon allotropes and explain in terms of its properties.

The valence shell electron-pair repulsion theory states that pairs of valence electrons around an atom in a molecule arrange themselves spatially, to get as far away from one another as possible. This determines the shape of the molecule. The repulsion between a lone pair and a bonded pair is greater than between bonded pairs; therefore, the presence of a lone pair of electrons will push the bonded pairs closer together. The table below summarises the shapes of molecules.

Total number of electron pairs	Arrangement of electron pairs	Bonding pairs	Lone pairs	Shape of molecule
2	Linear	2	0	Linear
3	Trigonal planar	3	0	Trigonal
4	Tetrahedral	4	0	Tetrahedral
		3	1	Pyramidal
		2	2	Bent
5	Trigonal bipyramidal	5	0	Trigonal bipyramidal
6	Octahedral	6	0	Octahedral

1 Draw Lewis dot structures then predict the shapes of the molecules below.

a Carbon dioxide

d Phosphorus pentachloride

Shape: _____

b Ammonia

e Boron trifluoride

Shape:_____

c Carbon tetrachloride

Shape: _____

f Hydrogen sulfide

Shape: _____

Shape: _____

2 Carbon dioxide and hydrogen sulfide are both molecules that are made up of three atoms; however, their shapes are very different. Explain why their shapes are different with reference to your answers in questions **1a** and **1f**.

_____

_____

_____

_____

_____

_____

_____

_____

_____

_____

_____

3 a Use the terms from the list provided to fill in the blanks. Some words may be used more than once or not at all.

bent	element	ozone
bonded	flat	parallel
buckminsterfullerene	forms	physical
chemical	gaseous	planar
compound	graphite	state
conduct	ionic	tetrahedral
covalently	linear	tetrahedrally
diamond	nanotubes	valence
different	oxygen	

Allotropes are forms of the one _____. In allotropes, atoms of the _____

are _____ to other atoms in _____ physical forms, in the

same _____. The element oxygen forms two types of allotropes, both in the _____

state. O_2 has a _____ shape, while O_3 has a _____ shape. The common

allotropes of carbon include: _____ , in which each carbon atom

is _____ bonded so there are no free electrons, and _____ , in which each

carbon atom is bonded to three other carbon atoms to form a _____ structure so there

are _____ electrons to _____ an electric current. In the latter structure, there ar

e _____ six-membered rings joined together in _____ layers. Other allotropes of

carbon include C_{60}, the _____ , and cylindrically shaped _____ .

b Identify the names of the allotropes of carbon shown.

i

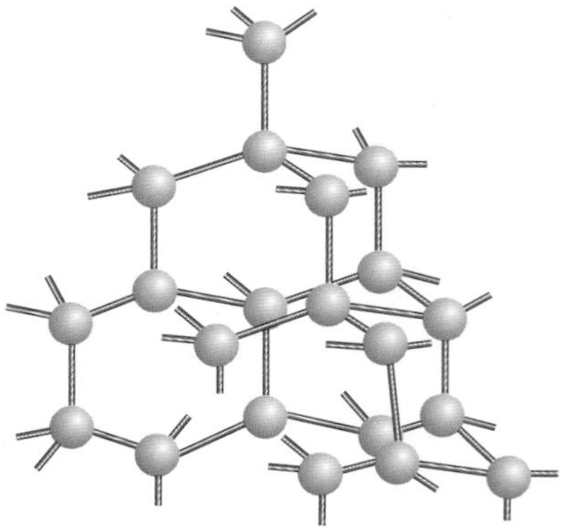

Name: _____

iii

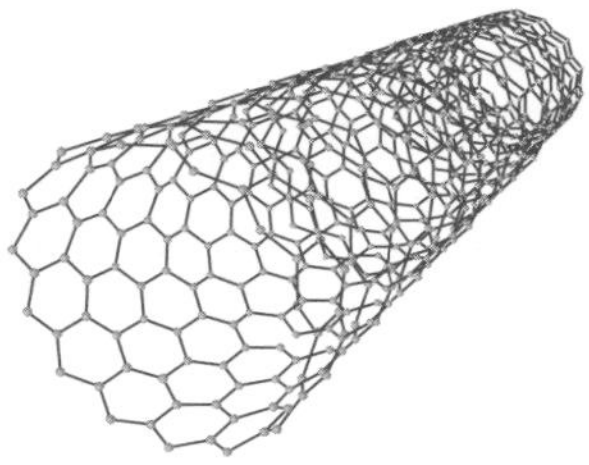

Name: _____

ii

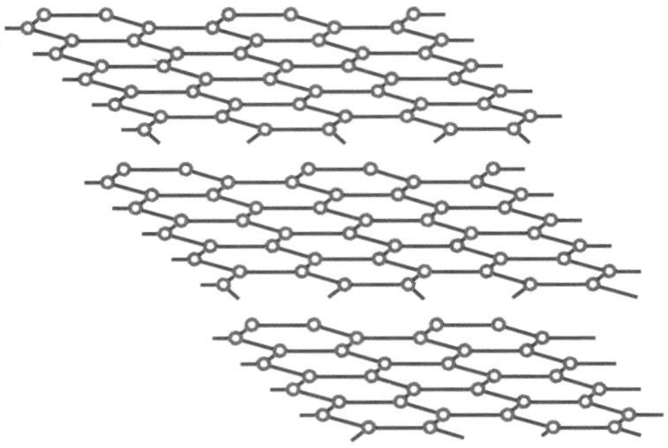

Name: _____

iv

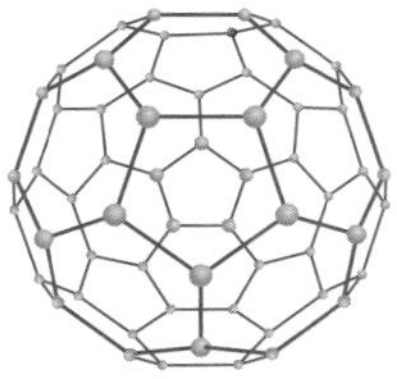

Name: _____

c Compare the hardness and electrical conductivity of structures **i**, **ii** and **iii** above. Be sure to include reasons for the properties.

_____

_____

_____

_____

_____

_____

_____

_____

d For each allotrope listed, write a use related to its property.

Allotrope	Use	Related property
Diamond		
Graphite		
Buckminsterfullerene		
Nanotube		

LEARNING GOALS

Describe intermolecular forces: dispersion forces, dipole–dipole forces and hydrogen bonds.

Identify types of intermolecular forces between molecules.

Relate boiling points to strength of intermolecular forces.

---

1 Complete the blanks using the words from the list. The words may be used more than once or not at all.

asymmetrical	less	shape	strongest
between	more	size	symmetrical
greater	non-polar	smaller	type
higher	polar	strength	weakest
larger	physical	stronger	

Intermolecular forces are attractive forces that exist _____ neighbouring molecules. All molecules

experience intermolecular forces that in turn determine their _____ properties such as melting

points and boiling points. The _____ the intermolecular force, the _____ the melting or

boiling point. The _____ of intermolecular forces present depend on whether the molecule is polar

or _____ and the elements that make up the molecule. Molecules may be _____ either

because they have non-polar bonds or because they have _____ bonds but are _____. The

intermolecular forces between these types of molecules are the _____ dispersion forces.

The _____ of the dispersion force depends on the _____ and _____ of the molecule;

i.e. the _____ the molecule and the _____ branched, the _____ the dispersion force.

Dipole–dipole forces are _____ than dispersion forces and exist between _____ molecules,

which have polar bonds and are _____. Hydrogen bonds are the _____ intermolecular force

and exist between molecules that have O-H, N-H or F-H.

2 Identify the type of intermolecular force between two of each of the same type of molecule below and draw diagrams to show where these forces occur.

a H_2O

c ICl

b HCl

d Cl_2

---

3 Pentane and 2,2-dimethylpropane have the same molecular formula, C_5H_{12}. Their structures are shown below. One has a boiling point of 10°C while the other has a boiling point of 36°C. Match the boiling points to the compounds and explain your answer.

Pentane

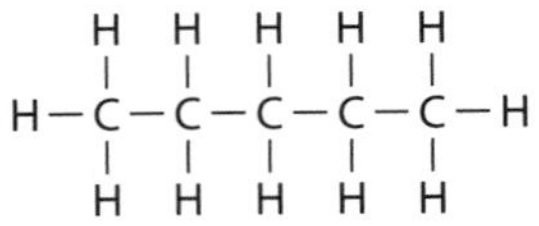

2,2-dimethylpropane

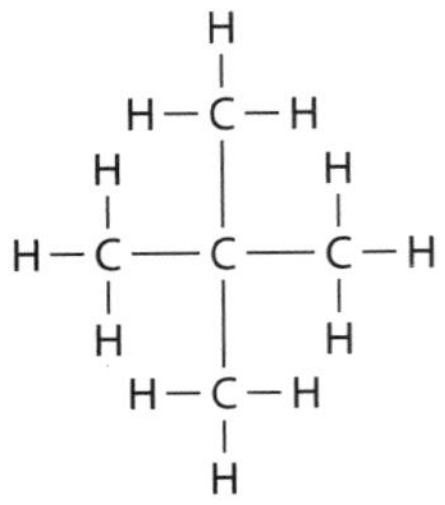

$$\begin{array}{cccccc} & H & H & H & H & H \\ & | & | & | & | & | \\ H- & C- & C- & C- & C- & C-H \\ & | & | & | & | & | \\ & H & H & H & H & H \end{array}$$

$$\begin{array}{ccc} & H & \\ & | & \\ & H-C-H & \\ & | & \\ H & | & H \\ | & | & | \\ H-C- & C- & C-H \\ | & | & | \\ H & | & H \\ & H-C-H & \\ & | & \\ & H & \end{array}$$

Boiling point: _____

Boiling point: _____

_____

_____

_____

_____

4 Label the intermolecular forces by drawing at least three neighbouring molecules of ethanamine, shown below.

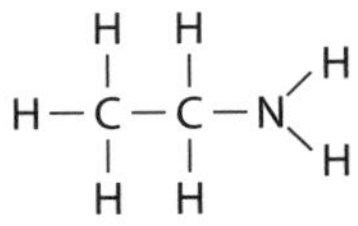

$$\begin{array}{ccc} H & H & \\ | & | & \quad H \\ H-C- & C- & N \diagup \\ | & | & \quad \diagdown H \\ H & H & \end{array}$$

Module one: Checking understanding

1 Match each term on the left with the appropriate description on the right by writing the number of the description next to the term.

A	Alpha ray		1	The solid that stays in the filter paper
B	Beta ray		2	Is the water when salt is added to it
C	Boiling point		3	Properties of elements vary periodically with their atomic numbers
D	Density		4	Has covalent network bonding
E	Diamond		5	The horizontal rows in the periodic table
F	Filtrate		6	Is a helium nucleus
G	Gamma ray		7	Is analysis by mass
H	Gravimetric analysis		8	Is defined as mass per unit volume
I	Group		9	Can be easily converted to a vapour
J	Non-metal		10	Are found in groups 3 to 12 in the periodic table
K	Period		11	Is generally dull and brittle
L	Periodic law		12	Silicon is an example of this type of element
M	Radioactivity		13	The liquid that passes through the filter paper
N	Residue		14	Can only be stopped by several centimetres of lead
O	Semi-metal		15	Is the spontaneous emission of radiation
P	Solvent		16	Is an electron
Q	Transition element		17	The lowest temperature at which a liquid changes to a gas
R	Valency		18	The vertical columns in the periodic table
S	Vanadium		19	Has 23 protons
T	Volatile		20	Is the combining power of an element

2 Match the name with the equipment shown.

condenser   evaporating basin   fractionating column   separating funnel
conical flask   filter funnel   round-bottom flask   tripod and gauze mat

a

Name: _____

b

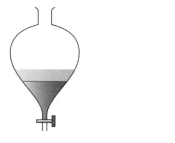

Name: _____

c

Name: _____

d

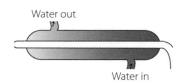

Name: _____

e

Name: _____

f

Name: _____

g

Name: _____

h

Name: _____

3 You have been provided with a mixture of X, Y and Z. Their properties are listed in the table. Outline a procedure for separating the components.

Substance	State at room temperature	Solubility of Y in	Boiling point (°C)	Polar/non-polar	Volatility
X	Liquid	Sparingly soluble	68	Non-polar	Volatile
Y	Solid		2230	Non-polar	Not volatile
Z	Liquid	Sparingly soluble	78	Polar and non-polar	Volatile

4 Aspartame is a chemical used in artificial sweetener. Its molecular formula is $C_{14}H_{18}N_2O_5$ and its structural formula is shown below.

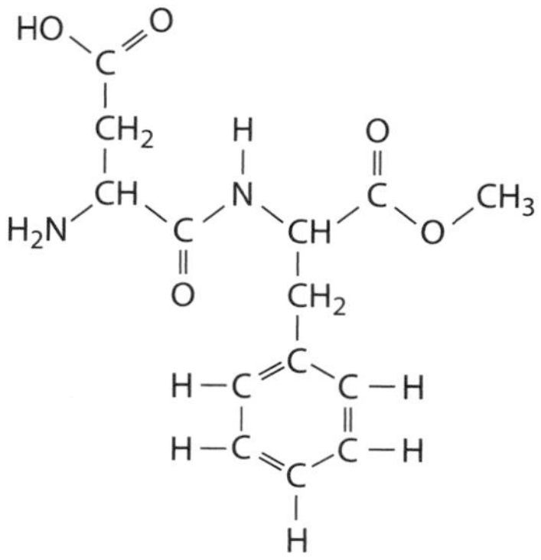

a State the names of the component elements in Aspartame.

b Calculate the percentage composition by weight of component elements, given the mass of the Aspartame and components shown in the table.

	Mass(g)
Aspartame	294.304
Carbon	14 × 12.01
Hydrogen	18 × 1.008
Oxygen	5 × 16.00
Nitrogen	2 × 14.01

_____

_____

_____

_____

_____

_____

_____

_____

_____

_____

5 Complete the table below.

Species	Electron configuration (*spdf* notation)
Carbon	
Chromium	
Copper	
Fluoride	
Lithium	
Magnesium	
Oxide ion	

6 a Identify the type of orbital shown by using the letters *s, p, d* and *f.*

i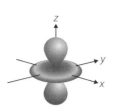

Name: _____

ii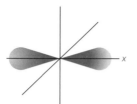

Name: _____

iii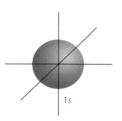

Name: _____

iv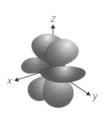

Name: _____

b State the maximum number of electrons one orbital can hold.

_____

7 Name the type of bonding and provide an example for each of the diagrams below.

a

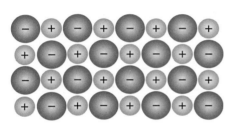

Type: _____

Example: _____

c

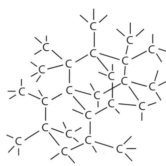

Type: _____

Example: _____

b

Type: _____

Example: _____

d

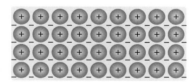

Type: _____

Example: _____

8 a Write nuclear equations to show uranium-238 undergoing alpha decay to form an element X. Then element X undergoes beta decay.

_____

_____

b Label the types of radiation in the diagram below.

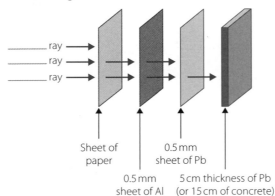

_____ ray ⟶

_____ ray ⟶

_____ ray ⟶

Sheet of
paper

0.5 mm
sheet of Pb

0.5 mm   5 cm thickness of Pb
sheet of Al   (or 15 cm of concrete)

9 a Explain how the coloured lines in emission spectra are obtained and why each element has its own unique spectrum.

_____

_____

_____

_____

_____

_____

_____

_____

_____

_____

b Describe how studying emission spectra of elements enabled the Bohr model to be further developed by Schrödinger.

_____

_____

_____

_____

_____

_____

_____

_____

10 a Explain the trend in atomic radius from the sodium atom to the chlorine atom.

_____

_____

_____

_____

_____

_____

b Compare the radius of the sodium ion with that of the oxide ion.

_____

_____

_____

_____

_____

_____

c Successive ionisation energies of two different elements, X and Y, are shown in the graphs below.

i Suggest the formula of the compound formed between X and Y, giving reasons for your response.

_____

_____

_____

_____

_____

_____

ii Suggest what elements X and Y could be.

_____

_____

_____

_____

11 The table shows the melting points of two different solid substances, X and Y, which are both soluble in water.

Substance	Melting point (°C)
X	146
Y	801

a   Suggest the type of bonding in X and Y.

_____

_____

_____

_____

_____

b   Suggest a physical test to identify the type of bonding in X and Y.

_____

_____

_____

_____

12 Distinguish between the terms 'allotropes' and 'isotopes' by using appropriate examples.

_____

_____

_____

_____

_____

_____

_____

Reviewing prior knowledge

1 a Mark each statement as True or False.

Statement	True or False?
A tomato ripening on a vine is an example of a chemical change.	
Hydrogen gas (H_2) is an example of a molecule, but not a compound.	
Salt, oxygen gas and iron(II) chloride solution are all pure substances.	
Saltwater is homogenous.	
Sedimentation cannot be used to separate wooden blocks from rocks.	
The electron configuration of nickel is $1s^2 2s^2 2p^6 3s^2 3p^6 4s^2 3d^7$.	
Diamond and graphite are allotropes of carbon.	
Copper salts burn with a red flame.	

b For any statements marked False above, write the correct statement below.

_____

_____

_____

_____

_____

2 Complete the table below.

Chemical name	Chemical formula	Chemical name	Chemical formula
	NaCl	Barium sulfate	
	MgO	Magnesium chloride	
	Na_2CO_3	Lithium phosphide	

3 Draw electron-dot structures to demonstrate how the electron arrangements of the atoms change following a chemical reaction.

a aluminium + chlorine → aluminium chloride

b hydrogen + sulfur → hydrogen sulfide

4 Some of the properties of calcium, bromine and calcium bromide are given in the table below.

Property	Calcium	Bromine	Calcium bromide
State at 25°C	Solid	Liquid	Solid
Colour	Silver	Reddish-brown	White
Melting point (°C)	842	−7.2	742
Boiling point (°C)	1484	59	1815
Density (g cm^{-3})	1.54	3.10	3.35

Use this information to explain why calcium bromide is said to be a compound and not a mixture of calcium and bromine.

_____

_____

_____

_____

_____

_____

5 A student performed an experiment to determine the percentage of water in a sample of hydrated copper(II) sulfate, $CuSO_4.5H_2O$.

The student's results are tabulated below.

Mass of crucible (g)	67.5
Initial mass of crucible + copper(II) sulfate before heating (g)	80.0
Mass of crucible + copper(II) sulfate after first heating (g)	78.4
Mass of crucible + copper(II) sulfate after second heating (g)	77.3
Mass of crucible + copper(II) sulfate after third heating (g)	76.1
Mass of crucible + copper(II) sulfate after fourth heating (g)	76.1

a Using the information provided by the student's results, write a suitable method by which the student could accurately determine the percentage of water in a sample of hydrated copper(II) sulfate.

_____

_____

_____

_____

_____

_____

_____

_____

_____

b Create a Risk Assessment for this experiment.

c Justify the need for the student to perform four heatings of the crucible and copper(II) sulfate.

_____

_____

d Calculate the theoretical and experimental percentage of water in the hydrated copper(II) sulfate ($CuSO_4.5H_2O$) sample used.

_____

_____

_____

_____

_____

HINT

Theoretical percentage is calculated using the chemical formula whereas experimental percentage uses the experimental results.

e Account for any discrepancy between the theoretical and experimental values.

_____

_____

_____

f Evaluate the reliability of this experiment and suggest a modification to the experiment that could improve reliability.

_____

_____

_____

_____

6 a Explain the trend in first ionisation energy across period 3 of the periodic table.

_____

_____

_____

_____

b Using the table below, identify the two exceptions to this trend, and explain why this is the case.

Element	Symbol	Atomic number	First ionisation energy /kJ mol^{-1}
Sodium	Na	11	496
Magnesium	Mg	12	738
Aluminium	Al	13	578
Silicon	Si	14	789
Phosphorus	P	15	1012
Sulfur	S	16	1000
Chlorine	Cl	17	1251
Argon	Ar	18	1521

_____

_____

_____

_____

_____

_____

HINT

To explain why, try drawing the electron configurations of the elements.

7 Define the term 'isotope'.

_____

_____

8 Label the parts of the isotopic notation.

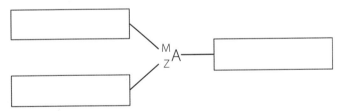

9 Complete the table below.

Isotope name	Isotopic notation	Atomic number	Number of protons	Number of neutrons
Helium-4				2
	$^{11}_{5}B$			
			19	18
Americium-241		95		
	$^{235}_{92}U$	92		

Chemical reactions and stoichiometry

 Applying the law of conservation of mass

 STUDENT BOOK
Pages 142–5

LEARNING GOALS

Design a valid experiment demonstrating the law of conservation of mass.

Process and evaluate data.

Apply the law of conservation of mass to balancing chemical equations.

1 State the law of conservation of mass.

_____

_____

2 A student wanted to demonstrate this law by conducting an experiment. A diagram of the experimental set-up is below.

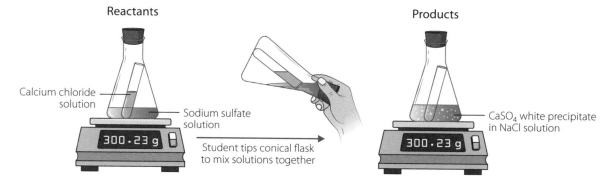

a Complete the following word equation:

calcium chloride + sodium sulfate → _____ + _____

b Write a hypothesis for this experiment.

_____

c Using the diagram above, write a suitable method that would enable this student to test their hypothesis.

_____

_____

_____

_____

_____

_____

_____

HINT

Remember to be specific and include the variables, controls and any safety requirements.

_____

_____

_____

d Write a conclusion for this experiment.

_____

_____

3 A student from another Chemistry class conducted a similar experiment on the law of conservation of mass by designing the following experimental method.

1 Measure 20 mL of vinegar. Pour the vinegar into a beaker. Weigh the beaker containing the vinegar and record its mass.

2 Weigh a 5.0 g sample of baking soda onto a clean watch glass. Record the mass.

3 Carefully pour the 5.0 g of baking soda into the beaker of vinegar. Record the mass of the beaker, vinegar and baking soda when the solution has stopped fizzing.

The chemical equation for this reaction is:

Baking soda + vinegar → sodium acetate + water + carbon dioxide

$$NaHCO_3(s) + CH_3COOH(aq) → NaCH_3COO(aq) + H_2O(l) + CO_2(g)$$

The student's results were recorded in the table below.

	Trial 1	Trial 2	Trial 3	Trial 4	Average (g)
Mass of beaker + vinegar (g)	231	242	235	234	
Mass of baking soda (g)	4.9	4.9	5.1	4.9	
Mass of beaker + vinegar + baking soda (g)	225	238	226	228	

a Identify any outliers present. If so, explain the potential sources of the error.

_____

_____

_____

_____

_____

_____

b Calculate the average results in the table.

c Explain whether the results for this experiment demonstrate the law of conservation of mass.

_____

_____

_____

_____

_____

4 a Write the following word equations as balanced chemical equations.

HINT

Remember diatomics and to include states.

   i iron + copper(II) chloride → copper + iron(II) chloride

_____

   ii zinc + hydrochloric acid → zinc chloride + hydrogen gas

_____

   iii fluorine gas + calcium → calcium fluoride

_____

   iv sodium hydroxide + sulfuric acid → sodium sulfate + water

_____

b For each of the reactions above, calculate the relative atomic mass or relative molecular mass for each reactant and product, and then calculate the total atomic/molecular mass of reactants and the total atomic/molecular mass of products.

_____

_____

_____

_____

_____

_____

_____

5 Using the equations in question **4a ii** and **iv** above as examples, explain why equations need to be balanced to reflect the law of conservation of mass.

_____

_____

_____

_____

_____

_____

_____

_____

_____

_____

_____

7 Mole concept

WS 7.1 Performing quantitative calculations

STUDENT BOOK
Pages 145–66

LEARNING GOALS

Define and distinguish between key quantitative terms.

Manipulate the formula $n = \dfrac{m}{MM}$ to suit the subject of the question and perform calculations.

Calculate the empirical and molecular formula of compounds.

---

1 Explain the differences between the terms.

a Relative atomic mass, relative molecular mass and relative formula mass

_____

_____

_____

_____

_____

_____

b Empirical formula and molecular formula

_____

_____

_____

c A mole and the molar mass of a substance

_____

_____

_____

2 a Mark each statement as True or False.

Statement	True or False
One mole of oxygen gas contains one mole of oxygen atoms.	
The mass of all atoms of an element is the same.	
The formula of an ionic compound gives the ratio of the atoms present in the compound.	
The Avogadro constant is equal to the number of atoms in exactly 12 g of the carbon-12 isotope.	
One mole of any substance would contain 6.03×10^{24} particles of that substance.	
A molecular formula gives the exact number of the different kinds of atoms in the molecule.	

b For each statement marked as False, write the correct statement below.

_____

_____

_____

3 Compare the steps involved in calculating the relative molecular mass of a molecule and the relative formula mass of an ionic compound.

HINT

Don't forget to include units in your final calculation.

_____

_____

_____

_____

_____

_____

_____

4 Complete the table with the correct formula that would be applied for each scenario, then complete the calculation.

Subject	Formula to use	Calculation
Determine the number of moles of gold in a nugget weighing 0.24 kg.		
Given there are 0.013 moles of sugar ($C_6H_{12}O_6$) in a cup of tea, determine the mass of the sugar cube used.		
A 2 mole sample of an unknown element weighs 53.96 g. Determine the element.		Element is:
A 100 g sample of a compound contains 62.1 g C, 27.7 g O and 10.3 g H. Determine the empirical formula of the compound.		Empirical formula is:

5 Consider the following reaction:

$$2H_2S(g) + 3O_2(g) \rightarrow 2H_2O(g) + 2SO_2(g)$$

Which of the following statements is incorrect?

A 2 mol of H_2S is required to produce 2 mol of H_2O.

B 6 mol of O_2 is required to produce 4 mol of SO_2.

C 1 mol of H_2S is required to produce 1 mol of SO_2.

D 0.75 mol of O_2 is required to produce 0.66 mol of H_2O.

6 The relative formula mass of $Fe_2(SO_4)_3$ is:

A 207.8   **B** 255.8   **C** 335.8   **D** 399.9

7 What is the mass of nitrogen in a 0.28 mol sample of aluminium nitrate, $Al(NO_3)_3$?

A 150.3 g   **B** 50.0 g   **C** 11.8 g   **D** 3.9 g

8 The number of moles of oxygen atoms in 22.8 g of sodium phosphate, Na_3PO_4 is:

A 0.56 moles   **B** 0.46 moles   **C** 0.35 moles   **D** 0.14 moles

9 Which of the following contains the largest mass of hydrogen?

A 44 g hydrogen sulfide (H_2S)

B 3.0 g hydrogen (H_2)

C 13 g of butane (C_4H_{10})

D 22 g methane (CH_4)

10 Complete the following calculations.

a Calculate the empirical formula of a molecule found to contain 54.5% carbon, 36.4% oxygen and 9.1% hydrogen.

b Calculate the molecular formula of a compound with a molecular mass of 135 g mol^{-1} found to contain 53.3% carbon, 35.6% oxygen and 11.1% hydrogen.

c A single displacement reaction occurs according to the following balanced equation:

$$Ag_2SO_4(aq) + Cu(s) \rightarrow CuSO_4(aq) + 2Ag(s)$$

If 0.412 mol of $CuSO_4$ is produced, how many mol of Ag(s) is produced?

d Look at the following balanced equation:

$$H_2SO_4(aq) + Mg(OH)_2(s) \rightarrow MgSO_4(aq) + 2H_2O(l)$$

If 3.46 g of $MgSO_4$ is produced, what is the mass of the water that forms?

_____

_____

_____

_____

11 Convert the following steps for calculating masses in a chemical reaction into a flow chart.

Step 1: Convert the mass of substance A into its corresponding number of moles using its molar mass.

Step 2: From the balanced chemical equation, calculate the number of moles of substance B from the number of moles of substance A using the appropriate mole ratio.

Step 3: Convert the number of moles of substance B into its mass using its molar mass.

LEARNING GOALS

Write a valid method to determine the empirical formula of varying forms of cobalt chloride.

Calculate the number of moles and determine the empirical formula from experimental data.

As a depth study, a student wanted to perform an experiment to determine the empirical formula of a hydrated sample of cobalt chloride ($CoCl_2$). During their research, the student noted that the dark red cobalt chloride underwent another colour change, to purple, before becoming the blue anhydrous form.

The student decided to try to determine the empirical formula of each substance.

1 Design a suitable method that could have been used by the student to determine the empirical formula of each of the cobalt chloride colours.

_____

_____

_____

_____

_____

_____

_____

_____

_____

_____

_____

2 The student performed the experiment. Recorded below is the mass of each colour of the cobalt chloride minus the weight of the crucible and lid.

Mass of red cobalt chloride (g)	15.81
Mass of violet cobalt chloride (g)	11.06
Mass of water lost (g)	
Mass of blue cobalt chloride (g)	8.66
Mass of water lost (g)	

a Calculate the mass of water lost between each colour transition of the cobalt chloride. Write your answers in the table above.

b Calculate the number of moles of blue anhydrous cobalt chloride.

_____

_____

c Calculate the number of moles of water lost as the violet cobalt chloride became anhydrous.

d Determine the empirical formula of the violet cobalt chloride ($CoCl_2.xH_2O$).

e Determine the empirical formula of the red cobalt chloride ($CoCl_2.xH_2O$).

HINT

Use the number of moles of blue anhydrous cobalt chloride calculated in part 2b to determine the ratio.

LEARNING GOALS

Write balanced chemical equations.

Use mole ratios to complete stoichiometry problems.

Review the law of conservation of mass.

Calculate experimental percentage mass, comparing this to theoretical mass to determine potential sources of error.

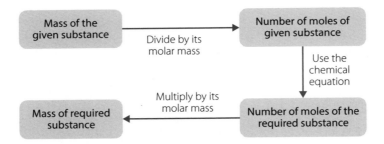

1 Plants use energy from the Sun to turn carbon dioxide and water into glucose and oxygen in a process called photosynthesis.

   a Write a balanced chemical equation for photosynthesis.

   _____

   b Determine the mass of glucose produced during a photosynthesis reaction in which a plant absorbs 10.0 moles of carbon dioxide.

   _____

   _____

   _____

   _____

   c Determine the volume of water in mL that the plant requires to react with the 10.0 moles of carbon dioxide.

   _____

HINT
1 mL of water weighs 1 g.

   _____

   _____

2 Gasification is an industrial chemical process where organic fuels such as coal are reacted with steam to produce methane (CH_4) and carbon monoxide. Assume the main component of coal that reacts is carbon (C).

   a Write a balanced chemical equation for this process.

   _____

   b Determine the mass of carbon required in order to produce 5.0 moles of methane.

   _____

   _____

   _____

   _____

c Often the coal used in this process contains other chemicals such as sulfur, resulting in the unwanted production of sulfur dioxide. Released into the atmosphere, this may result in the production of acid rain according to the following equation:

$$2SO_2(g) + O_2(g) + 2H_2O(l) \rightarrow 2H_2SO_4(l)$$

An environmental scientist determined 78.0 moles of sulfuric acid were produced by a company. Determine the mass of sulfur dioxide released.

_____

_____

_____

_____

d Every 10.0 kg of coal (C) used in the gasification process produces 7.0 g of sulfur dioxide. Determine the mass of coal reacted to produce the 78.0 moles of sulfuric acid found by the scientist.

_____

_____

_____

_____

_____

_____

3 Alka Seltzer tablets are often taken to relieve the symptoms of acid reflux. They contain sodium bicarbonate ($NaHCO_3$) that reacts with stomach acid to produce sodium chloride, water and carbon dioxide. A student added a tablet to a beaker of hydrochloric acid and recorded the following results.

Mass of the tablet (g)	3.23
Mass of beaker (g)	329.62
Mass of beaker + HCl (g)	364.37
Mass of beaker + contents after bubbling stopped (g)	363.46
Mass of CO_2 released (g)	

a Write the balanced chemical equation for the reaction of sodium bicarbonate with hydrochloric acid.

_____

b Complete the table determining the mass of CO_2 released.

c Using the law of conservation of mass, explain why the mass of carbon dioxide can be determined in this manner.

_____

_____

_____

_____

_____

d Using the data from the table and your balanced equation, determine the experimental mass of sodium bicarbonate in the Alka Seltzer tablet.

_____

_____

_____

_____

e The Alka Seltzer package lists the mass of sodium bicarbonate per tablet to be 1.90 g. Use this mass, and the experimental mass calculated above to determine the percentage experimental error.

_____

_____

_____

f Provide suggestions as to why the mass of sodium bicarbonate calculated from the experimental data did not equal the theoretical mass provided by the packaging.

_____

_____

_____

_____

_____

_____

_____

_____

_____

_____

_____

_____

_____

_____

_____

WS 7.4 Introducing limiting reagents

Understand the concept of limiting reagents through analogy.

Balance chemical equations.

Use mole ratios to solve problems.

Determine the limiting reagent in chemical reactions.

Determine the excess of a chemical in a reaction.

1 Limiting reagents can be explained using the following analogy.

A club sausage sizzle is being run at the local soccer field. For the breakfast shift, they are serving bacon and egg sandwiches that require 2 pieces of bread, 3 pieces of bacon and 1 egg. They begin the day with 300 pieces of bread, 440 pieces of bacon and 200 eggs.

a Given the starting number of groceries, how many sandwiches can be made based on the number of:

i pieces of bread?

_____

ii pieces of bacon?

_____

iii eggs?

_____

b Identify the limiting reagent in making a bacon and egg sandwich.

_____

c Calculate the excess quantities of the remaining reagents.

_____

_____

_____

d The goal of the club sausage sizzle was to raise $1000 for the club. Provided each bacon and egg sandwich was sold for $2.50, determine the number of starting reagents (bread, bacon and eggs) required to make enough sandwiches to reach the target.

_____

_____

_____

2 A 1 mole sample of magnesium reacts with 2 moles of hydrochloric acid to produce 1 mole of magnesium chloride and 1 mole of hydrogen gas.

a Write the balanced equation, including states.

_____

b If a 2 mole sample of magnesium was used, how many moles of hydrochloric acid would be required?

_____

c If a 4 mole sample of magnesium was used, how many moles of hydrogen gas would be produced?

_____

d If 1 mole of magnesium is added to just 1 mole of hydrochloric acid:

   i how many moles of the magnesium would remain unreacted?

_____

   ii how many moles of hydrogen gas would be produced?

_____

e How many grams of magnesium would be required to react with 2 moles of hydrochloric acid?

_____

f How many grams of hydrogen would be produced if a 48.62 gram sample of magnesium was completely reacted?

_____

g If a 0.25 mole sample of hydrochloric acid was used, how many grams of magnesium would be required for a complete reaction?

_____

_____

h **i** Identify the limiting reagent if a 2 gram sample of magnesium was added to 0.2 moles of hydrochloric acid.

_____

_____

_____

   ii Determine the number of moles of excess reactant remaining.

_____

3 A 4 mole sample of iron reacts with 3 moles of oxygen gas to produce 2 moles of iron(III) oxide.

a Write the balanced formula, including states.

_____

b If an 8 mole sample of iron was used, how many moles of oxygen would be required?

_____

c If a 2 mole sample of iron was used, how many moles of iron(III) oxide would be produced?

_____

d If a 4 mole sample of iron is exposed to just 1.5 moles of oxygen:

   i how many moles of the iron would remain unreacted?

_____

   ii how many moles of iron(III) oxide would be produced?

_____

e How many grams of iron would be required to react with 3 moles of oxygen?

_____

_____

f How many grams of iron(III) oxide would be produced if a 111.7 gram sample of iron was completely reacted?

_____

_____

_____

g If 0.8 moles of oxygen gas was used, how many grams of iron would be required for a complete reaction?

_____

_____

h **i** Identify the limiting reagent if a 5.0 gram iron nail was exposed to 0.0090 moles of oxygen from the air.

_____

_____

_____

_____

_____

ii Determine the number of moles of excess reactant remaining.

_____

_____

iii Determine the mass of the rust (Fe_2O_3) produced.

_____

_____

_____

8 Concentration and molarity

WS 8.1 Solving concentration and molarity problems

STUDENT BOOK
Pages 173–82

LEARNING GOALS

Define the terms 'concentration' and 'molarity'.

Calculate the concentration of a solution.

Calculate the molarity of a solution.

Determine the number of ions in a solution.

Write balanced chemical equations.

Determine the percentage composition of a chemical.

1 What is the difference between concentration and molarity?

_____

_____

_____

2 Diesel engines produce dangerous nitrogen oxides as a by-product of combustion. To ensure these oxides do not cause an environmental concern, a diesel exhaust fluid (DEF), which is made by mixing a percentage of urea (CH_4N_2O) with deionised water, is added to the combustion process to lower nitrogen oxide emissions.

a Calculate the number of grams of urea required to make 1 L of a $2.0 \, mol \, L^{-1}$ solution of DEF.

_____

_____

b Determine the mass of urea in 100 mL of a 5% (w/v) DEF solution.

_____

c Calculate the mass of urea needed to make a 12% (w/w) DEF solution using 250 mL of deionised water.

_____

_____

d Determine the molarity of a DEF solution made using 4.5 g of urea in a 120.0 mL volumetric flask.

_____

_____

_____

3 A volumetric flask contains 250 mL of a $0.0125 \, mol \, L^{-1}$ magnesium chloride solution.

a Calculate the number of moles of magnesium chloride in the flask.

_____

_____

b Calculate the moles of chloride ions in the flask.

c Calculate the total number of ions in the flask.

d Calculate the mass of magnesium chloride used to make the solution in the flask.

4 The active ingredient in bleach is the compound sodium hypochlorite (NaClO). Determine the concentration of sodium hypochlorite, expressed as %(w/w), in a $0.25\,mol\,L^{-1}$ sample of bleach.

5 A 3.8 g antacid tablet contains 75% magnesium hydroxide, $Mg(OH)_2$, by mass.

a Write a balanced chemical equation for solid magnesium hydroxide reacting with hydrochloric acid to produce magnesium chloride and water.

b Determine the number of moles of $Mg(OH)_2$ in the tablet.

c Calculate the number of moles of HCl required to react with the $Mg(OH)_2$.

d What volume of $0.30\,mol\,L^{-1}$ HCl could the antacid tablet neutralise?

6 For safety precautions, sulfuric acid is often transported at high concentrations such as $14.0 \, mol \, L^{-1}$ rather than in dilute forms.

    a Calculate the mass of sulfuric acid present in 25.0 L of this concentrated solution.

_____

_____

    b If the density of sulfuric acid is $1.840 \, g \, mL^{-1}$, calculate the percentage composition (% w/w) of this solution.

_____

_____

_____

    c Calculate the final molarity of a sulfuric acid solution when 100 mL of $0.2 \, mol \, L^{-1}$ H_2SO_4 is mixed with 200 mL of $1.2 \, mol \, L^{-1}$ H_2SO_4.

_____

_____

_____

_____

_____

Performing dilutions

LEARNING GOALS

Model concentrated and dilute solutions.

Calculate the volume of water required to make a dilution.

Determine the most accurate piece of equipment to perform a dilution.

1 In the beakers below, draw a model to represent a concentrated and a dilute solution.

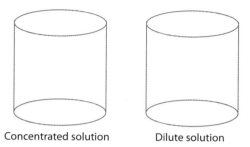

Concentrated solution   Dilute solution

2 Serial dilutions are a simple technique for reducing the concentration of a solution in a systematic manner.

a The diagram below shows the steps involved in creating a serial dilution where the concentration is reduced by a factor of 10 each time. Beginning with $10\,mol\,L^{-1}$ LiCl solution, which has a strong orange colour, annotate the diagram to show:

i the concentration of the solution in each test tube

ii the volumes of LiCl required to be transferred

iii the volumes of water added

iv the comparative colour of each test tube.

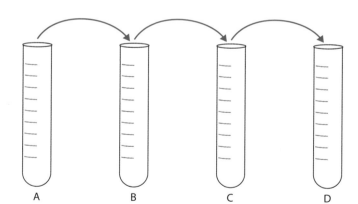

A   B   C   D

b Determine the volume of solution required from test tube B in order to make a $0.72\,mol\,L^{-1}$ solution in a new 10 mL test tube.

_____

_____

_____

_____

c A student stated that they could make a $0.02\,\text{mol}\,\text{L}^{-1}$ solution from test tube D. Their math calculation is below:

$$V_1 = \frac{c_2 V_2}{c_1}$$

$$= \frac{0.02 \times 10}{0.01} = 20\ \text{mL}$$

Evaluate this student's statement.

_____

_____

_____

_____

_____

_____

3 a What volume of water should a chemist use to dilute $133\,\text{mL}$ of a $7.90\,\text{mol}\,\text{L}^{-1}\,CuCl_2$ solution so that $51.5\,\text{mL}$ of the diluted solution contains $4.49\,\text{g}\,CuCl_2$?

_____

_____

_____

_____

_____

_____

_____

b The chemist did not have the most appropriate equipment on hand. Based on the equipment list below, determine the most accurate piece of equipment the chemist could use. Justify your choice.

- $100\,\text{mL}$ measuring cylinder with $\pm0.5\,\text{mL}$ uncertainty
- $50\,\text{mL}$ burette with $\pm0.2\,\text{mL}$ uncertainty
- $500\,\text{mL}$ beaker with $\pm1\,\text{mL}$ uncertainty

_____

_____

_____

_____

_____

_____

_____

_____

$$V_1 = \frac{c_2 V_2}{c_1}$$

WS 9.1 Introducing kinetic theory

LEARNING GOALS

Review the spacing, arrangement and movement of particles in a solid, liquid and gas.

Label a water phase change graph.

Model the behaviour of an ideal gas.

Interpret data from a graph.

Describe the trends in data.

1 Complete the table summarising the spacing, arrangement and movement of particles for each state of matter.

	Solid	Liquid	Gas
Spacing			
Arrangement			
Movement of particles			
Diagram			

2 Place labels for the following at the most relevant point or section on the graph below.

 ▶ Solid, liquid, gas

 ▶ Boiling point, melting point

 ▶ Phase changes (melting, evaporation, freezing, condensation)

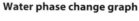

Water phase change graph

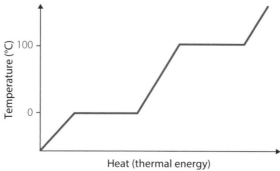

Heat (thermal energy)

3 The kinetic theory of gases assumes that:

- all gases consist of molecules (except noble gases, which consist of atoms)
- gas molecules move very fast
- gas molecules move in straight lines until they collide with other molecules
- all collisions are perfectly elastic, so no energy is lost
- the molecules are all moving in random directions
- the distance between gas molecules is very large compared to the size of the molecules
- intermolecular forces are negligible, so gas molecules don't clump together at all.

Create a three-dimensional visual model of the behaviour of an ideal gas, illustrating as many of the assumptions listed above as you can. Around the diagram, add labels indicating which assumptions are being modelled.

4 The graph below shows the distribution of speeds for different gases at the same temperature.

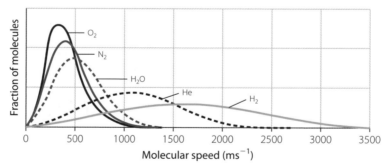

a Why do the particles of each particular gas have a range of speeds at the same temperature?

_____

b What is the average speed of the molecules of each gas?

_____

_____

_____

c   What is the relative molecular or atomic mass of each gas?

_____

_____

_____

d   What is the relationship between average speed and molecular/atomic mass?

_____

_____

_____

e   Describe the trend between the mass of particles and the variations in speed.

_____

_____

_____

_____

WS 9.2 Calculating molar volume of gases

1 Represent the following steps in performing molar gas calculations as a flow chart, assuming standard conditions:

Step 1: Convert the volume gas A into its corresponding number of moles using molar volume.

Step 2: Using a balanced chemical equation, calculate the number of moles of gas B from the number of moles of gas A.

Step 3: Convert the number of moles of gas B into its mass using its molar mass.

2 Four litres of CO_2 contains the same number of moles as 4 L of O_2, yet 4 g of MgO doesn't contain the same number of moles as 4 g of BaO.

Evaluate this statement.

_____

_____

_____

_____

_____

_____

_____

3 Three balloons are lined up on a fence, filled with three different gases. Using the information from the table below, complete the diagram by drawing the size of the balloons to represent their volume.

Gases	H$_2$	He	CH$_4$
Pressure (kPa)	100	100	100
Temperature (°C)	0	0	0
Mass (g)	2.016	4.003	16.042

4 A person was blowing up a beachball.

a Explain why the beachball expands with every breath the person blows into the ball.

_____

_____

_____

_____

b How will the size of the beachball blown up on a cold day (0°C) compare to the same ball blown up on a warm day (25°C)?

_____

_____

_____

5 When magnesium metal reacts with dilute sulfuric acid, it produces hydrogen gas and an aqueous solution of magnesium sulfate, according to the following equation:

$$Mg(s) + H_2SO_4(aq) \rightarrow MgSO_4(aq) + H_2(g)$$

An experiment was set up as below.

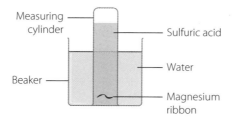

a Write a suitable aim for this experiment.

_____

_____

b Describe how three variables could be controlled in this experiment.

_____

_____

_____

_____

_____

c Write a method for this experiment.

_____

_____

_____

_____

_____

_____

_____

_____

_____

_____

_____

d　**i**　If a 2.0 g sample of magnesium was allowed to completely react in this experiment at 25°C and 100 kPa, calculate the volume of H_2 produced.

_____

_____

_____

_____

　ii　Explain the main experimental flaw if a 2.0 g sample of magnesium was reacted within a 100 mL measuring cylinder.

_____

_____

e If the volume of H_2 recorded was 50.5 cm^3, calculate the mass of the magnesium remaining from the 2.0 g sample. Assume the experiment was conducted at 25°C and 100 kPa.

_____

> **HINT**
>
> $1\,cm^3 = 1\,mL$

_____

_____

_____

_____

LEARNING GOALS

Apply an understanding of Boyle's law to scuba diving.

Determine the volume of gas/es within a scuba tank.

Apply an understanding of Gay-Lussac's law to scuba diving.

Calculate the pressure within a scuba tank.

Calculate the temperature within a scuba tank.

Scuba diving is a recreational activity in which a person uses a Self-Contained Underwater Breathing Apparatus (SCUBA) to supply them with oxygen, allowing extended durations of sea diving. The air we breathe on the surface comprises 78% nitrogen, 21% oxygen and 1% other gases. The average air pressure at sea level is 1 atmosphere (unit = atm); however, in water, the pressure increases 1 atm for every 10 m in depth the diver descends.

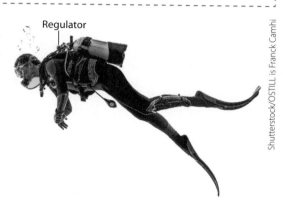
Regulator

Shutterstock/OSTILL is Franck Camhi

1 Assuming a full scuba tank contains 500 L of air, calculate the volume of oxygen and nitrogen within the tank.

_____

_____

HINT

Determine the percentages of oxygen and nitrogen that would make up the 500 L of air based on their atmospheric composition.

2 Calculate the pressure experienced by a diver, in kPa, at 35 m below sea level.

_____

_____

HINT

1 atm = 101.325 kPa

3 Suggest why diving 35 m below sea level affects the human body more than being 35 m above sea level.

_____

_____

_____

4 One of the fundamental rules of scuba diving is not to hold your breath as you ascend to the surface, but rather continuously breathe out. Explain the need for this rule using Boyle's law.

_____

_____

_____

_____

_____

_____

_____

_____

To prevent divers' lungs from collapsing as they descend to great depths, the compressed air from the scuba tank is passed through a device called a regulator, which reduces the pressure from the scuba tank to match the ambient pressure of the diver.

HINT

1 psi = 0.068 atm = 6.89 kPa

Another vital piece of equipment is the scuba tank. The air within it is compressed to around 3000 psi (pound-force per square inch) and the containers themselves are made from steel or aluminium to withstand the high-pressure environments. The capacity of scuba tanks is often stated as the amount of breathing gas the tank will hold at its maximum rated service pressure. 'Amount' is typically stated as the equivalent volume at a pressure of 1.0 atm.

5   If an 80.0 ft^3 tank is filled to its rated pressure of 3000 psi, the gas in the tank would have a volume of 80.0 ft^3 at a pressure of 1.0 atm. If the tidal volume (the amount of air the diver breathes in or out during normal respiration) is 500 mL, how many breaths should they expect from their tank at a depth of 35 m?

HINT

1 ft^3 = 28.32 L

6   When tanks are rapidly filled with air, the temperature of the tank can increase up to 60°C. Using Gay-Lussac's law, explain what potential problem this poses for divers.

7   When a diver fills an 80.0 ft^3 scuba tank to 3000 psi, the temperature of the scuba tank reaches 55.0°C. After 2 hours the tank has cooled to an ambient temperature of 19.0°C. What is the actual pressure in the tank?

HINT

Temperature should be in K.

As temperature plays such a vital role in pressure, scuba tanks are fitted with release valves called 'burst disks' that will release excess pressure if the tank exceeds the maximum capacity. Most burst disks are designed to rupture when pressure rises 15% above the rated 3000 psi.

8   A diver left a full scuba tank in the boot of their car while they washed down the boat after an overnight dive. It was a cool 19.0°C when he picked up the tank but as it became mid-morning, the ambient temperature inside the car was increasing. Calculate the temperature that will cause the burst disks to rupture on the tank.

LEARNING GOALS

Identify and graphically represent the gas laws from real-life examples.

Compare the properties of an ideal gas to a real gas.

Make calculations using the ideal gas equation.

Interpret graphical data to answer questions on the compression factors of a gas.

1 Using the information provided from the real-life example, identify the most relevant gas law and draw a graphical representation showing the relationship between the variables of this law.

Gas law	Graphical representation	Real-life example
		As humans inhale, the volume of the lungs increases along with the molar quantity of oxygen.
		Pressurised containers have safety labels warning that the container must be kept away from fire and stored in a cool environment or risk explosion.
		The 'bends' can occur when a scuba diver ascends to the surface too rapidly. The sudden decrease in pressure causes gases in the blood to expand and may cause organ damage or death.
		Helium balloons left out overnight will shrink due to the cold morning air.

2 Explain why the volume of a gas is larger when measured in real-life compared to the value determined by the ideal gas equation calculation.

_____

_____

_____

3 Use the ideal gas equation to complete the following questions.

  a A 25.8 L sealed canister is heated to 37°C, at a pressure of 135 kPa. Calculate the number of moles of gas within the container.

_____

_____

_____

_____

  b A container holding 2.64 moles of gas at a temperature of 8.20°C exerts a pressure of 233 kPa. Calculate the volume of the container.

_____

_____

_____

_____

  c A 12.8 L canister containing carbon dioxide gas was stored at a temperature of 25.8°C and 99.3 kPa. Calculate the mass of the gas in the canister.

_____

_____

_____

_____

4 An ideal gas is one for which the ideal gas equation, $PV = nRT$, applies at all temperatures and pressures.

In reality, no gas is perfectly ideal. If a gas behaved in an ideal way, PV would always be equal to nRT. If PV and nRT are the same, then if you divide PV by nRT the answer would, of course, be 1. The expression $\dfrac{PV}{nRT}$ is called the compression factor.

If we plot the compression factor as the pressure and/or temperature of a gas are varied, the resulting curve should be a straight, horizontal line showing a compression factor of 1. But if a gas does not always act in an ideal way, the curve will not be a straight, horizontal line.

The graph below plots the compression factor of nitrogen gas as the pressure is increased at different temperatures. (Note: 1 bar = 100 kPa)

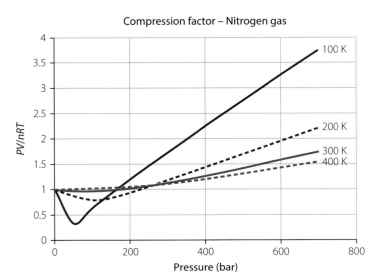

a For a temperature of 300 K, describe what happens to the compression factor as the pressure increases up to 200 bar at that temperature, and suggest a reason why this might be the case.

_____

_____

b Describe what happens to the compression factor as the pressure increases above 200 bar at room temperature. Suggest why this might happen.

_____

_____

_____

_____

c What happens to the compression factor at temperatures below room temperature as the pressure increases up to 200 bar? Suggest why this might happen.

_____

_____

_____

_____

_____

d What happens to the compression factor at temperatures above room temperature as the pressure increases up to 200 bar?

_____

_____

e What happens to the compression factor at all temperatures as the pressure increases above 200 bar? Suggest why that might be happening.

_____

_____

_____

_____

f What happens to the compression factor at pressures above 200 bar as the temperature increases?

_____

_____

_____

g Based on the information from the graph, write a generalisation about the conditions under which a gas no longer behaves like an ideal gas.

_____

_____

Module two: Checking understanding

1 Match each term on the left with the appropriate description on the right by writing the number of the description next to the term.

A	Aqueous solution		1	Mass of a mole of a substance: can be used for both elements and compounds
B	Avogadro constant, N_A		2	Average mass of the atoms present in the naturally occurring element relative to the mass of an atom of carbon-12 isotope taken as exactly 12
C	Law of conservation of mass		3	Solution that contains a relatively high amount of solute, say, greater than about 50 g/100 mL
D	Limiting reagents		4	Law stating that, when measured at the same temperature and pressure, equal volumes of different gases contain the same number of molecules
E	Molar mass		5	Law stating that matter can be neither created nor destroyed, but can be changed from one form to another
F	Mole		6	Study of quantitative aspects of chemical formula and equations
G	Percentage composition		7	Solution in which the concentration is accurately known
H	Relative atomic mass		8	Amount of solute present in a specified amount of solvent or solution
I	Relative formula mass		9	Number of moles of solute per litre of solution
J	Relative molecular mass		10	For a fixed quantity of gas $P_2V_2/T_2 = P_1V_1/T_1$, where P_1, V_1, T_1 and P_2, V_2, T_2 are the initial and final pressures, volumes and absolute temperatures
K	Stoichiometry		11	Solution in which the solvent is water
L	Concentrated solution		12	Law stating that, for a given quantity of gas at a constant temperature, the product of its volume, V, and its pressure, P, is constant; i.e. $PV =$ constant
M	Dilute solution		13	Mass of a molecule of a compound relative to the mass of an atom of the carbon-12 isotope taken as exactly 12
N	Concentration		14	Volume occupied by one mole of a gas; it is the same for all gases (at the same temperature and pressure)
O	Molarity		15	$PV = nRT$
P	Standard solution		16	Quantity that contains as many elementary units (e.g. atoms, ions or molecules) as there are atoms in exactly 12 g of carbon-12 isotope
Q	Avogadro's law		17	Law stating that, when measured at constant temperature and pressure, the volumes of gases taking part in a chemical reaction show simple, whole number ratios to one another
R	Boyle's law		18	Constant in the ideal gas law with a value of 8.314 $J^{-1}K^{-1}$/mol
S	Charles's law		19	Number of atoms in exactly 12 g of the carbon-12 isotope
T	Combined gas law		20	(By weight) ratio by mass in which elements are present
U	Gay-Lussac's law of combining volumes		21	Unit temperature on the absolute scale of temperature
V	Gay-Lussac's pressure, temperature law		22	A solution that contains a relatively low amount of the solute, say, less than about 10 or 20 g/100 mL (less than about 10 or 20%(w/w))
W	Ideal gas law		23	Law stating that, for a fixed sample of gas at constant volume, pressure increases linearly with temperature
X	Kelvin		24	Reactant that is all used up in a chemical reaction, thus limiting the amount of product that can be formed
Y	Molar volume of a gas		25	Mass of a unit of the compound as represented by its formula, relative to the mass of carbon-12 atom taken as exactly 12
Z	Universal gas constant, R		26	Law stating that, at constant pressure, the volume of a fixed quantity of gas is proportional to its absolute (or Kelvin) temperature

2 Identify the stoichiometric ratio between phosphorus pentachloride and water in the following unbalanced reaction:

$$PCl_5 + H_2O \rightarrow H_3PO_4 + HCl$$

A 1:4   **B** 4:1   **C** 2:1   **D** 1:1

3 Nine molecules of oxygen react with excess hydrogen to produce water. How many molecules of water are produced?

A 1   **B** 4.5   **C** 9   **D** 18

4 When methane combusts in oxygen the following reaction occurs:

$$CH_4(g) + 2O_2(g) \rightarrow CO_2(g) + 2H_2O(l)$$

If the combustion of 4.8 g of methane produces 17.6 g of carbon dioxide and 14.4 g water, determine the mass of oxygen required in the reaction.

A 12.8 g   **B** 27.2 g   **C** 32.0 g   **D** 40.0 g

5 The relative formula mass of $Fe_2(SO_4)_3$ is:

A 104   **B** 208   **C** 320   **D** 400

6 A student weighs a sample of sodium phosphate (Na_3PO_4), then calculates that the sample contains 0.15 mol of the compound. Determine the mass of the sodium present in this sample.

A 1.9 g   **B** 5.9 g   **C** 10.3 g   **D** 28.0 g

7 The quantity of alcohol in mixer drinks is often expressed as %(v/v). Which of the following drinks contains the largest volume of alcohol?

A 150 mL 11% wine   **B** 450 mL 5% beer   **C** 30 mL 37% spirits   **D** 345 mL 4% cider

8 A saltwater solution has a concentration of 1.50 $mol\,L^{-1}$. What volume of water is added to 500 mL of this solution to change the concentration to 0.8 $mol\,L^{-1}$?

A 133.3 mL   **B** 266.6 mL   **C** 437.5 mL   **D** 937.5 mL

9 A balloon contains 3.5 L of air at 25°C and 101.3 kPa. If the pressure dropped to 92.8 kPa, while the temperature remained constant, determine the new volume.

A 3.3 L   **B** 3.8 L   **C** 87.2 L   **D** 91.7 L

10 A 0.8 mol sample of chlorine has a volume of 4.2 L. Assuming temperature and pressure remain constant, determine the volume occupied by 2.1 mol.

A 0.4 L   **B** 1.6 L   **C** 9.0 L   **D** 11.0 L

11 Air behaves most like an ideal gas when at:

A absolute zero.

B high pressure.

C high altitude.

D high temperature.

Reviewing prior knowledge

1 Match each term on the left with the appropriate description on the right by writing the number of the description next to the term.

A	Anion		1	Attraction between cations and anions in an ionic compound
B	Ionisation energy		2	Attraction between delocalised valence electrons and metal cations
C	Cation		3	Attraction between protons and electrons of participating atoms
D	Chemical bond		4	Attraction between shared electrons and nuclei of atoms involved
E	Compound		5	Change in which no new substance is formed
F	Covalent bonding		6	Charged atom or group of atoms
G	Molecular compound		7	Compound that is made up of positive ions and negative ions
H	Valence electrons		8	Electrostatic force of attraction between molecules in close proximity
I	Delocalised electrons		9	Measure of the ability of an atom of an element to attract bonding electrons towards itself in compounds
J	Valency		10	Detached electrons that are free to move about within the metal
K	Polyatomic ion		11	Number of single bonds with a H atom that the atom could form
L	Intermolecular bond		12	Discrete molecule in which atoms are joined by covalent bonding
M	Intramolecular bond		13	Electrons in the outer shell of an atom in its ground state
N	Ion		14	Electrostatic force of attraction that holds atoms together within a molecule
O	Ionic bonding		15	Pure substance composed of different atoms bonded together
P	Ionic compound		16	Ions that consist of two or more atoms strongly covalently bonded
Q	Metallic bonding		17	Negatively charged atom or group of atoms
R	Electronegativity		18	Energy required to remove an electron from a gaseous atom of an element
S	Chemical change		19	Process in which at least one new compound is formed
T	Physical change		20	Positively charged atom or group of atoms

2 Give the valency of the following elements.

a Br _____   d P _____   g S _____   j Al _____

b Ca _____   e Ar _____   h Na _____   k He _____

c B _____   f H _____   i Si _____   l Cl _____

3 Write balanced chemical equations for the following reactions.

a Magnesium reacts with oxygen gas.

_____

b Calcium carbonate decomposes to form calcium oxide and carbon dioxide gas.

_____

c Methane gas (CH_4) completely combusts to form carbon dioxide gas and water vapour.

_____

d Nitric acid reacts with zinc to form zinc nitrate and hydrogen gas.

_____

e Sodium carbonate reacts with hydrochloric acid to form sodium chloride, carbon dioxide gas and water.

_____

4 Explain why the law of conservation of mass is important for balancing chemical equations.

_____

_____

_____

_____

_____

_____

_____

5 a Explain what is meant by the term 'rate of reaction'.

_____

_____

b List four factors that affect the rate of a reaction and describe how each changes the rate.

_____

_____

_____

_____

_____

_____

_____

_____

WS 10.1 Identifying reaction type

STUDENT BOOK
Pages 223–35

LEARNING GOALS

Identify different reaction types.

Predict the products of a reaction.

Write balanced chemical equations.

1 a The following table summarises the different types of reactions. Unfortunately, the definitions have been mixed up. Use a line to match the reaction type with the correct definition.

Type of reaction	Definition	Equation
1 Combustion	A Two or more substances join to form a new substance.	$CH_4(g) + O_2(g) \rightarrow$
2 Decomposition	B A solid forms when two solutions are mixed.	$CuCO_3(s) \rightarrow$
3 Synthesis/direct combination	C A compound is broken down into two other substances.	$Fe(s) + S(s) \rightarrow$
4 Precipitation	D A substance reacts with oxygen at a temperature well above room temperature.	$AgNO_3(aq) + NaCl(aq) \rightarrow$

b Complete the equation for each of the reaction types in the table above.

2 Below are diagrams that represent three of the above reaction types. For each diagram, name the reaction type represented and give reasons for your answer.

a

_____

_____

_____

b

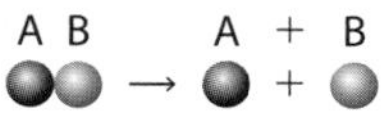

A B   A + B

⬤⬤ → ⬤ + ⬤

_____

_____

_____

c

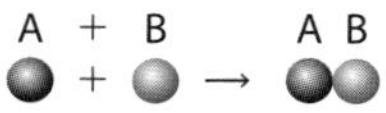

A + B   A B

⬤ + ⬤ → ⬤⬤

_____

_____

_____

3 Explain why it is difficult to draw a similar representation for the reaction type that is missing from question **2**. Give two different reactions to support your explanation.

_____

_____

_____

_____

_____

_____

4 Use your knowledge of the following reaction types to complete the equations below.

   a Decomposition

      i $Al_2O_3(s) \rightarrow$

      ii $H_2O(l) \rightarrow$

   b Synthesis

      i $Na(s) \ + \ Br_2(g) \rightarrow$

      ii $H_2(g) \ + \ Cl_2(g) \rightarrow$

   c Combustion

      i $Fe(s) \ + \ O_2(g) \rightarrow$

      ii $C_2H_4(g) \ + \ O_2(g) \rightarrow$

   d Precipitation

      i $Na_2CO_3(aq) \ + \ Al(NO_3)_3(aq) \rightarrow$

      ii $Ba(OH)_2(aq) \ + \ Cu(NO_3)_2(aq) \rightarrow$

LEARNING GOALS

Process precipitation investigation data.

Identify the products of precipitation reactions.

Derive generalisations from investigation data.

Analyse data to identify inconsistencies, suggest reasons and solutions.

Construct balanced neutral species and net ionic equations for precipitation investigation reactions.

Use data to identify unknown samples.

A class conducted a series of investigations to observe examples of precipitation reactions. Different groups were given different combinations of anions and cations.

The results of the groups are given below where ✓ shows there was a precipitate and ✗ means there was no precipitate.

Group 1

	Cl^-	SO_4^{2-}	CO_3^{2-}	OH^-
Ca^{2+}	✗	✓	✓	✗
Mg^{2+}	✗	✗	✓	✓
Cu^{2+}	✗	✗	✓	✓
Zn^{2+}	✗	✗	✓	✓
Ag^+	✓	✓	✓	✓

Group 3

	NO_3^-	SO_4^{2-}	CO_3^{2-}	O^{2-}
Na^+	✗	✗	✗	✗
Mg^{2+}	✗	✗	✓	✓
Cu^{2+}	✗	✗	✓	✓
NH_4^+	✗	✗	✗	✗
Ba^{2+}	✗	✓	✓	✗

Group 2

	Cl^-	SO_4^{2-}	NO_3^-	OH^-
Ca^{2+}	✗	✓	✗	✗
NH_4^+	✗	✗	✗	✗
Pb^{2+}	✓	✓	✗	✓
Ba^{2+}	✗	✓	✗	✗
Ag^+	✗	✓	✗	✗

Group 4

	Cl^-	NO_3^-	CO_3^{2-}	O^{2-}
Ca^{2+}	✗	✗	✓	✗
Pb^{2+}	✓	✗	✓	✓
Na^+	✗	✗	✗	✗
Zn^{2+}	✗	✗	✓	✓
Ag^+	✓	✗	✓	✓

1 a i Collate the class data given above.

ii Identify any differences in results and suggest reasons for the differences.

_____

_____

_____

_____

_____

iii For any different results, suggest what should be done to check which group is correct.

_____

_____

_____

_____

_____

b Write generalisations based on the data.

_____

_____

_____

_____

_____

_____

_____

_____

_____

3 Write balanced neutral species equations for the reactions between each of the following:

a $Na_2CO_3(aq) + Cu(NO_3)_2(aq) \rightarrow$ _____

b $Ca(NO_3)_2(aq) + Na_2SO_4(aq) \rightarrow$ _____

c $AgNO_3(aq) + NaCl(aq) \rightarrow$ _____

4 Write the net ionic equation for each of the following reactions:

a magnesium nitrate and sodium hydroxide

_____

b lead nitrate and ammonium oxide

_____

_____

_____

c copper sulfate and barium chloride

_____

5 The students were given three unknown samples labelled A, B and C. They were told the samples were magnesium sulfate, lead nitrate and barium hydroxide. Suggest a series of tests that could be conducted to identify each sample and give the possible results for each test.

_____
_____
_____
_____
_____
_____
_____
_____

WS 10.3 Evaluating an acid–carbonate investigation

STUDENT BOOK
Pages 242–3

LEARNING GOALS

Assess risk in an investigation.

Justify dependent, independent and controlled variables.

Evaluate experimental controls for reliable data collection.

Determine if an investigation requires modification.

Construct balanced equations for acid–carbonate reactions.

Identify the salt produced when an acid reacts with a carbonate.

Students conducted the following investigation.

AIM

To investigate the reactions of acids with carbonates.

MATERIALS

- 2 or 3 marble chips (calcium carbonate)
- Other carbonates; e.g. sodium carbonate, sodium hydrogen carbonate, potassium carbonate, copper carbonate
- Dilute (1 mol l^{-1}) hydrochloric acid (HCl)
- Limewater solution
- Distilled water
- Conical flask
- Measuring cylinder
- 2 test tubes
- Stopper fitted with U tube
- Stand and clamp

METHOD

1 Set up the apparatus as shown in the diagram below.

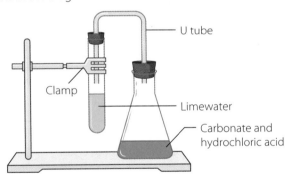

2 Add 2 or 3 pieces of marble to the flask. Add 10 mL of dilute hydrochloric acid and quickly place the stopper into the top of the flask.

3 One-third fill a test tube with limewater and allow the gas to bubble into the limewater. Record your observations. If the limewater turns milky it indicates carbon dioxide is present.

4 Feel the flask containing the acid and carbonate. Record your observation.

5 Rinse out the conical flask with distilled water. Then repeat steps 2–4 with the other carbonates.

RESULTS

The limewater turned cloudy for all samples of carbonate used.

Use the information provided above and your knowledge of chemistry to answer the following questions.

1 Complete the following risk assessment table.

What are the risks in doing this experiment?	How can you manage risks to stay safe?
Hydrochloric acid is corrosive.	
Limewater is dangerous and can seriously burn the eyes.	
Some chemicals may not be able to be disposed of down the sink.	

2 a Identify the independent and dependent variables. Give reasons for your selections.

_____

_____

_____

_____

b Describe how three other variables will be controlled.

_____

_____

_____

_____

3 Consider your answer to question 2 and the given aim of the investigation.
a Evaluate the use of experimental controls for reliable data collection.

_____

_____

_____

_____

_____

_____

_____

_____

9780170449564

b Determine whether the experiment needs to be modified and, if so, suggest modifications.

_____

_____

_____

_____

_____

_____

_____

_____

_____

_____

4 **a** Write a balanced equation for the reaction between HCl and calcium carbonate.

_____

b Write a general equation for the reaction between an acid and a carbonate.

_____

5 **a** Give the name of the salt produced by the reaction in question **4a**.

_____

b If sulfuric acid was used instead of hydrochloric acid, name the salt that would have been produced.

_____

c What salt would have been produced if nitric acid and copper carbonate were used for the reaction?

_____

Predicting reactions of metals

WS 11.1 Analysing metal reactivity data

STUDENT BOOK
Pages 250–4

LEARNING GOALS

Interpret secondary data to compare reactivity of metals.

Interpret secondary data to identify patterns of reactivity of metals.

Construct a metal activity series using secondary data.

Write balanced equations for reactions of metals.

Assess risk in an investigation.

Students used the results of experiments and second-hand sources to compile the following table of reactivity of different metals.

Element	K	Na	Li	Ca	Mg	Al	Zn	Fe	Pb	Cu	Ag	Au
Reaction with air at room temperature	Reacts rapidly				Reacts slowly				No reaction			
Reaction when heated in oxygen	Burns vigorously		Burns strongly	Burns quite fast	Burns vigorously	Goes white	Goes yellow	Moderate reaction	Reacts slowly	Surface blackens	No reaction	
Reaction in water at room temperature	Reacts very violently	Reacts violently	Fast reaction	Reacts readily	Reacts very slowly	No observable reaction						
Reaction when heated in steam	Explosive		Very violent	Violent	Burns rapidly	Little reaction	Moderate reaction	Reacts slowly	No observable reaction			
Reaction in dilute HCl and H_2SO_4	Explosive reaction		Violent reaction	Reactive with HCl	Reactive with both	Slow reaction	Quite reactive	Reacts slowly	No observable reaction			

Use the information in the table to assist you in answering the questions below.

1 Compare the reactions of metals with air and when heated in oxygen. Propose two reasons for the differences.

_____

_____

_____

_____

_____

_____

_____

2 Some metals react when heated in oxygen until all the metal has reacted, while other metals react to form an oxide coating, which prevents further reaction.

    a Using the data in the table, name one metal that undergoes complete reaction when heated in oxygen and write an equation to represent this reaction.

_____

    b Using the data in the table, name two metals that form an oxide coating and write an equation to represent this reaction.

_____

3 Some metals react violently with water. Suggest what steps a student should take to minimise risks associated with performing an experiment to test the reactivity of these metals with water.

4 **a** Use the information in the results table provided previously to identify on the periodic table below where the most active and least active metals are located.

Li																	
Na	Mg										Al						
K	Ca					Fe			Cu	Zn							
									Ag								
									Au		Pb						

    b Make a summary statement about how activity of metals changes from left to right across the table.

_____

_____

_____

5 **a** Write down a reactivity series for the metals listed in the results table.

_____

    b Which metal does not appear to react as strongly as it ought to, according to the reactivity series?

_____

6 a Write a general word equation for the reaction of reactive metals with oxygen gas.

_____

b Write balanced equations for the reactions of the following metals with oxygen.
  i Sodium

_____

  ii Calcium

_____

  iii Iron

_____

7 a Write a general word equation for the reaction of reactive metals with water.

_____

b Write balanced equations for the reactions of the following metals with water at room temperature.
  i Potassium

_____

  ii Magnesium

_____

  iii Zinc

_____

8 a Write a general word equation for the reaction of reactive metals with acids.

_____

b Write balanced equations for the reactions of the following metals with dilute hydrochloric acid.
  i Lithium

_____

  ii Copper

_____

  iii Magnesium

_____

WS 11.2 Interpreting metal activity data

LEARNING GOALS

Explain reasons for experimental procedures.

Interpret experimental data.

Identify and explain relationships in experimental data.

Write balanced equations.

Determine species being oxidised and reduced.

Analyse correlations between metal activity and atomic radius; metal activity and electronegativity.

Students were given information about an experiment that had been conducted.

AIM

To compare the reactivity of different metals.

MATERIALS

- Small pieces of copper, lead, iron, magnesium, zinc, aluminium
- 20 mL of each of the following 0.1 mol L^{-1} solutions:
  - Copper(II) sulfate ($CuSO_4$)
  - Iron(II) sulfate ($FeSO_4$)
  - Lead(II) nitrate ($Pb(NO_3)_2$)
  - Zinc sulfate ($ZnSO_4$)
  - Magnesium sulfate ($MgSO_4$)
  - Silver nitrate ($AgNO_3$)
  - Aluminium nitrate ($Al(NO_3)_3$)
- 36 test tubes
- Test tube racks
- Sandpaper

RISK ASSESSMENT

What are the risks in doing this investigation?	How can you manage these risks to stay safe?
Eye irritation	Wear safety glasses.
Lead nitrate is toxic	Avoid contact with skin. Wash with plenty of water if contact occurs.
Silver nitrate will stain skin	Avoid contact.
Water pollution via waste products	Dispose of silver nitrate, copper sulfate and lead nitrate solutions by pouring them into the provided waste bottle. Do not dispose of them down the sink.

METHOD

1 Place six (6) test tubes in a rack and add 3 mL of each of the solutions into separate test tubes.
2 Add to each of the test tubes a small piece of a metal that has been first cleaned with sandpaper.
  (Do not add the same metal to its own solution; e.g. do not place magnesium in a magnesium sulfate solution.)
3 Observe and record results. If no change is visible within five (5) minutes, write 'NR' (no reaction).
4 Dispose of the solutions as directed. Do not pour pieces of metal down the sink.
5 Wash the pieces of metal carefully if they are being reused.
6 Repeat the above procedure with all other metals.

Use the information provided above and your knowledge of chemistry to answer the following questions.

1 Explain why:

a the silver nitrate, copper sulfate and lead nitrate solutions should not be poured down the sink.

_____

_____

b the pieces of metal had to be cleaned with sandpaper before being used.

_____

_____

c a piece of metal did not need to be placed in a solution of its own ion.

_____

The experiment was conducted by a student and the following results were obtained.

R means a reaction occurred and NR means no reaction.

Solution/metal	Cu	Fe	Pb	Zn	Mg	Al
Copper(II) sulfate		R	R	R	R	R
Iron(II) sulfate	NR		NR	R	R	R
Lead(II) nitrate	NR	R		R	R	R
Zinc sulfate	NR	NR	NR		R	R
Magnesium sulfate	NR	NR	NR	NR		NR
Aluminium nitrate	NR	NR	NR	NR	R	
Silver nitrate	R	R	R	R	R	R

2 Write balanced equations for the reaction between the substances listed below and explain the species being oxidised and the species being reduced.

a Zinc and copper(II) sulfate

_____

_____

_____

b Iron and lead(II) nitrate

_____

_____

_____

3 Use the results of the experiment to list the six metals in order from most reactive to least reactive, explaining your reasoning.

_____

_____

_____

_____

4 Consider the possible reactivity of silver relative to the other metals. Where would you place silver in the activity list? Why?

_____

_____

_____

5 A different student performed this investigation and they suggested that aluminium was the least reactive metal of the ones being tested. This result was not supported by any other students. Explain possible error/s in their method that could have led to this result.

_____

_____

_____

6 The following table gives the atomic radius and electronegativity of the metals investigated.

Metal	Atomic radius (pm)*	Electronegativity (Pauling scale)
Mg	160	1.31
Al	143	1.61
Fe	124	1.83
Cu	128	1.9
Zn	133	1.65
Ag	144	1.93
Pb	175	1.8

* This is one half of the shortest interatomic distance in the crystalline metal.

a i Use the data in the table above to list the metals in order from largest to smallest atomic radius.

_____

ii Compare the list with your answer to question **3** and identify whether there is any correlation. If there is, suggest a reason this might occur.

_____

_____

_____

_____

_____

b i Use the data in the table above to list the metals in order from highest to lowest electronegativity.

_____

ii Compare the list with your answer to question **3** and identify whether there is any correlation. If there is, suggest a reason this might occur.

_____

_____

_____

_____

_____

_____

_____

LEARNING GOALS

Use definitions of oxidation and reduction and assess the correctness of statements.

Determine the oxidation number of elements in a compound.

Identify species being oxidised and species being reduced.

Identify and assess oxidant and reductant.

Write balanced half and full equations.

Explain an application of redox reactions.

1 Classify each of the following statements as true (T) or false (F). For those statements that are false, rewrite the statement so it is correct.

 a Reduction is the gain of electrons by a substance.

 _____

 b When copper loses two electrons to form Cu^{2+}, it is reduced.

 _____

 c An oxidation reaction is always accompanied by a reduction reaction.

 _____

 d A more active metal will displace a less active metal from a solution of its ions.

 _____

 e For positive monatomic ions, the oxidation state is always $+1$.

 _____

 f The oxidation number of manganese in MnO_2 is $+1$.

 _____

 g For any redox reaction, the number of electrons lost must be equal to the number of electrons gained.

 _____

 h In the reaction $CuO(s) + H_2(g) \rightarrow Cu(s) + H_2O(l)$, copper goes from an oxidation state of $+1$ to 0.

 _____

 i For any neutral compound, the sum of the oxidation numbers of the atoms in the molecule must equal zero.

 _____

2 Give the oxidation number of each of the following elements in the compounds list below.

 a Potassium bromide _____

 b Magnesium _____

 c Aluminium oxide _____

 d Iron(II) chloride _____

 e Iodine _____

 f Iron(III) chloride _____

3 For each of the following reactions:

   i identify the species that is oxidised and the species that is reduced

   ii name the oxidant and reductant.

   a $Cl_2 + 2HBr \rightarrow 2HCl + Br_2$

   b $I_2O_5 + 3CO \rightarrow I_2 + 3CO_2$

   c $6Mn^{3+} + I^- + 6OH^- \rightarrow 6Mn^{2+} + IO_3^- + 3H_2O$

4 Explain why metals can act as reducers, but not as oxidisers.

5 In acidified aqueous solutions, chloride ions, $Cl^-(aq)$, react with permanganate ions to form chlorine, Cl_2, but do not react with dichromate ions. Explain which is the more powerful oxidant in aqueous solution: $MnO_4^-(aq)$ or $Cr_2O_7^{2-}(aq)$?

6 Write balanced half-equations, and use these to construct a balanced overall equation, for the reactions represented by the following skeleton equations.

   a $Na(s) + Cl_2(g) \rightarrow NaCl(s)$

   b $Fe^{2+}(aq) + MnO_4^-(aq) \rightarrow Fe^{3+}(aq) + Mn^{2+}(aq)$

   c $Cu(s) + NO_3^-(aq) \rightarrow Cu^{2+}(aq) + NO_2(g)$

d $MnO_2(s) + Cl^-(aq) \rightarrow Mn^{2+}(aq) + Cl_2(g)$

_____

_____

_____

7 A student read that silver jewellery can be cleaned by putting it in an aluminium pot containing hot salty water. They conducted an experiment to see if this was correct and found that it worked. They wanted to know why, so they did some research and found out the following information:

 ▸ Silver becomes tarnished because of a coating of silver sulfide, Ag_2S.

 ▸ Al is a more active metal than Ag.

 ▸ The half-equations for the reactions are:
  $Ag_2S(s) + 2e^- \rightarrow 2Ag(s) + S^{2-}(aq)$
  $2Al(s) + 3H_2O(l) \rightarrow Al_2O_3(s) + 6H^+(aq) + 6e^-$

Use the information to answer the following questions.

a Name the species being oxidised and give the change in oxidation number.

_____

b Name the species being reduced and give the change in oxidation number.

_____

c Explain why this is a better cleaning method than using an abrasive polish to clean the jewellery.

_____

_____

_____

LEARNING GOALS

Develop an inquiry question and hypothesis.

Identify the use of variables and evaluate experimental controls.

Represent results using an appropriate format.

Analyse and identify relationships in experimental results.

Modify an investigation.

Determine oxidation number and identify species being oxidised and reduced.

Write balanced full and half-equations.

Iron nails left outside tend to rapidly rust (or corrode) unless they have been galvanised. A student designed an investigation to identify whether all four factors (air, moisture, light and warmth) are required for iron nails to rust. Five test tubes were set up and the results after 48 hours are shown below.

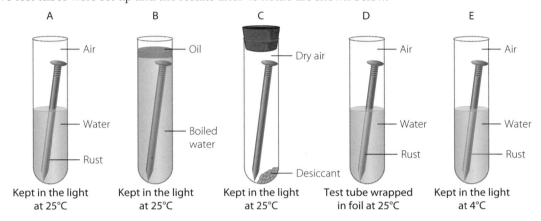

A	B	C	D	E
Air, Water, Rust	Oil, Boiled water	Dry air, Desiccant	Air, Water, Rust	Air, Water, Rust
Kept in the light at 25°C	Kept in the light at 25°C	Kept in the light at 25°C	Test tube wrapped in foil at 25°C	Kept in the light at 4°C

Answer the following questions related to the investigation described above.

1 Write a research question for this investigation.

_____

_____

2 Write a hypothesis for this investigation.

_____

_____

3 a State the purpose of each of test tubes B, C, D and E.

_____

_____

_____

_____

_____

_____

_____

_____

b Explain how the set up achieved the purpose for each of the test tubes in part **a**.

_____

_____

_____

_____

_____

_____

_____

_____

_____

_____

4 a What was the purpose of test tube A?

_____

_____

b What does test tube A by itself show?

_____

_____

c What does test tube A by itself not prove?

_____

_____

5 a Summarise the results of the investigation in a table.

b What condition/s are definitely required for rusting to occur? Justify your statement.

_____

_____

c Which condition/s are definitely **not** required for rusting to occur? Justify your statement.

_____

_____

6 a Explain whether you can assume from the experiment that rusting would occur at freezing temperatures.

_____

_____

b Describe how you could modify the investigation to check whether rusting would occur at freezing temperatures.

_____

_____

_____

_____

7 In the investigation described above, iron metal (Fe) rusts to eventually form hydrated Fe_2O_3.

a Identify the change in oxidation number of:

i Fe _____

ii O _____

b Justify the species that has been oxidised and the species that has been reduced.

_____

_____

c Write a half-equation for the species that has been oxidised.

_____

d In your answer to question **5b**, you identified the condition/s that must be present for rusting to occur. Identify which, if any, conditions do not appear in the final product of the reaction and explain why.

_____

_____

_____

_____

8 Rusting actually occurs through a series of reactions. Initially the iron atoms lose electrons to form Fe^{2+} ions. The electrons produced by this reaction are then used in the following reduction:

$$O_2(g) + 2H_2O(l) + 4e^- \rightarrow 4OH^-(aq)$$

Using the information above, write an equation for the first step in the rusting process.

_____

LEARNING GOALS

Explain how and why variables should be controlled.

Analyse the purpose of equipment in an experiment.

Represent and construct galvanic cells.

Analyse experimental data and compare with theoretical data.

Write half and full equations.

Calculate standard reduction potential using tabulated values.

Identify possible experimental errors and suggest improvements.

Students were provided with the following materials. They were told the aim of the investigation they were to perform was to measure the differences in voltages when different metals were used in constructing galvanic cells.

MATERIALS

- $1\,mol\,L^{-1}$ zinc nitrate $(Zn(NO_3)_2)$
- $1\,mol\,L^{-1}$ copper(II) nitrate $(Cu(NO_3)_2)$
- $1\,mol\,L^{-1}$ lead(II) nitrate $(Pb(NO_3)_2)$
- $1\,mol\,L^{-1}$ iron(II) sulfate $(FeSO_4)$
- $1\,mol\,L^{-1}$ magnesium sulfate $(MgSO_4)$
- Saturated potassium nitrate solution
- Strips of zinc, copper, iron, lead and magnesium metal

- $5 \times 100\,mL$ beakers
- $1 \times 250\,mL$ beaker
- Stirring rod
- Strips of filter paper about 1 cm wide and 10 cm long
- Voltmeter
- Connecting wires
- Fine sandpaper or steel wool

Students set up five galvanic half-cells and then used different combinations of the half-cells to make different galvanic cells.

1 a Explain how variables should be controlled.

_____

_____

_____

b Explain why these variables should be controlled.

_____

_____

2 a What is the purpose of the strips of filter paper and potassium nitrate solution?

_____

b What would be the result if this wasn't used?

_____

_____

c Explain how it works.

_____

_____

_____

_____

3 Draw a labelled diagram to show what one of the completed galvanic cells should look like.

4 Identify any issues the students should be aware of when constructing the galvanic cells and actions they could take.

_____

_____

_____

_____

_____

The students constructed the following results table and entered the voltage they obtained for each galvanic cell combination. Polarity indicates which electrode was positive and which was negative.

Beaker 1/ beaker 2	Polarity	Voltage (V)	Anode reaction	Cathode reaction	Theory value	% error
Zinc/copper	±	1.02				
Zinc/iron	±	0.2				
Zinc/lead	±	0.6				
Zinc/ magnesium	∓	0.9				
Iron/copper	±	0.64				
Iron/lead	±	0.26				
Iron/ magnesium	∓	1.41				
Lead/ magnesium	∓	1.52				
Lead/copper	±	0.51				
Magnesium/ copper	±	2.03				

5 Complete the anode and cathode reactions in the table.

6 **a** Write the overall reaction for the galvanic cell that gave:

  i the lowest voltage

  _____

  ii the highest voltage.

  _____

 b Compare the positions of each of the metals in parts **ai** and **aii** with those in a table of standard reduction potentials. What do you conclude?

  _____

  _____

  _____

  _____

  _____

7 Compare the voltages of the cells with the positions of the metals in the activity series and identify any relationship.

  _____

  _____

  _____

  _____

  _____

8 **a** Use a table of standard reduction potentials to calculate the $E°$ value for each galvanic cell combination.

  _____

  _____

  _____

  _____

  _____

  _____

  _____

 b Calculate the percentage error in the experimental results compared to the theoretical values.

  _____

  _____

  _____

  _____

  _____

  _____

c Compare the theoretical value with the value obtained in the experiment. Suggest reasons for any differences and possible improvements to the method.

_____

_____

_____

_____

_____

_____

_____

_____

_____

_____

_____

9 Draw a diagram of the electrochemical cell made by combining a Mg, Mg^{2+} electrode with a Hg^{2+}, Hg electrode with a platinum wire as an inert electrode.

Use your knowledge of the reactivity of metals to determine which metal will reduce (displace) which metal ion. Hence, write equations for the half-reactions and the overall reaction that occur in this cell. Indicate which electrode is positive and show the direction of electron flow and migration of ions. Identify the anode and cathode.

1 Use a table of standard electrode potentials to answer the following.

   a If a galvanic cell by made by connecting a Ag^+, Ag half-cell to a Cu^{2+}, Cu half-cell, which half-cell will be the cathode, and which the anode? What will be the cell potential?

   _____

   _____

   b If a galvanic cell is made by connecting a Cu^{2+}, Cu half-cell to a $PbSO_4$, Pb half-cell, which half-cell will be the cathode, and which the anode? What will be the cell potential?

   _____

   _____

   c If a galvanic cell is made by connecting a $PbSO_4$, Pb half-cell to a Zn^{2+}, Zn half-cell, which half-cell will be the cathode, and which the anode? What will be the cell potential?

   _____

   _____

   d What relationship exists among the cell potentials calculated in parts **a**, **b**, and **c** with the cell potential of a galvanic cell made by connecting a Ag^+, Ag half-cell to a Zn^{2+}, Zn half-cell?

   _____

   _____

   _____

   _____

2 Chemists decided to use a reference value other than setting the standard electrode potential for the H^+, H_2 half-cell to be zero. Explain whether the relative abilities of species to compete for electrons would be different.

   _____

   _____

3 Suppose that chemists had decided to use as a reference that the standard electrode potential for the H^+, H_2 half-cell is 5.00 V.

   a On this scale, what would be the standard electrode potentials for:

   i the Ag^+, Ag half-cell? _____

   ii the Cu^{2+}, Cu half-cell? _____

   iii the $PbSO_4$, Pb half-cell? _____

   iv the Zn^{2+}, Zn(s) half-cell? _____

b Using the values on this scale, recalculate the answers to question **1**. Comment on your answers here compared with those for question **1**.

_____

_____

_____

_____

4 The scale used is based on setting the standard electrode potential of the H^+, H_2 half-cell to be 0.00. Does this mean that the H^+, H_2 half-cell has zero oxidising power (or zero ability to compete for electrons)? Explain your answer.

_____

_____

_____

_____

5 For each of the following reactions, justify whether it will be spontaneous.

a $Sn + 2Ag^+ \rightarrow Sn^{2+} + 2Ag$ _____

b $2Ag^+ + Ni \rightarrow 2Ag + Ni^{2+}$ _____

c $3Fe + 2Al^{3+} \rightarrow 3Fe^{2+} + 2Al$ _____

d $Fe + Br_2 \rightarrow Fe^{2+} + 2Br^-$ _____

6 a Which is the strongest reductant out of Pb, Al, Fe Cu, and which is the weakest? Use the table of standard electrode potentials to justify your answer.

_____

_____

_____

_____

_____

_____

b Which is the strongest oxidant out of K^+, Ag^+, Al^{3+} and Ni^{2+}, and which is the weakest? Use the table of standard electrode potentials to justify your answer.

_____

_____

_____

_____

_____

_____

_____

7 The following cells were set up: however, one of the half-cells was unidentified. Use the standard voltage of the cell and a table of standard reduction potentials to:

    i calculate the voltage of the unknown half-cell

    ii identify the electrode of the unknown half-cell

    iii suggest a possible electrolyte for the unknown half-cell.

  a a copper, copper ion cathode half-cell is connected with a salt bridge to X, X ion half-cell

    $E^{\ominus}{}_{total} = +0.60$ V

_____

_____

_____

_____

_____

_____

_____

_____

_____

  b an aluminium, aluminium ion anode half-cell is connected to Y, Y ion half-cell

    $E^{\ominus}{}_{total} = +0.50$ V

_____

_____

_____

_____

_____

_____

_____

_____

_____

_____

INQUIRY QUESTION: WHAT AFFECTS THE RATE OF A CHEMICAL REACTION?

WS 12.1 Interpreting experimental results

STUDENT BOOK
Pages 298–307

LEARNING GOALS

Write a research question and hypothesis.

Identify experimental controls to ensure reliable data.

Evaluate an investigation set-up.

Analyse experimental data.

Write balanced chemical equations.

Make predictions related to data.

In an investigation, 2 g of marble chips (calcium carbonate) were placed in excess dilute hydrochloric acid, and the gas produced in the reaction was measured using the apparatus shown.

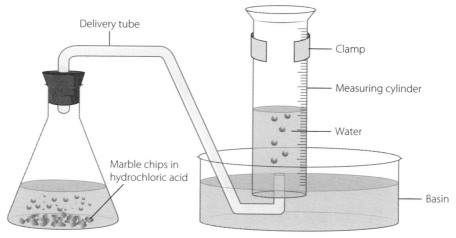

Delivery tube

Clamp

Measuring cylinder

Water

Marble chips in hydrochloric acid

Basin

The apparatus was used to measure the gas produced each minute over a 10-minute period, then repeated using 2 g of powdered $CaCO_3$ and fresh dilute excess HCl.

1 a Write a research question for this investigation.

_____

_____

b Write a hypothesis for this investigation.

_____

_____

_____

c Describe the independent and dependent variables.

_____

_____

_____

d Explain what needs to be done to ensure valid and reliable data is collected.

_____

_____

_____

_____

2 Explain why the measuring cylinder was inverted and initially full of water.

_____

_____

3 Describe how you would confirm that the gas produced is carbon dioxide and write a balanced chemical equation for this reaction.

_____

_____

4 Write a balanced chemical equation for the reaction between the marble chips and hydrochloric acid.

_____

Following are the results of the investigation.

Time (min)	Volume gas (mL)	
	Using chips	**Using powder**
0	0	0
1	20	70
2	75	130
3	130	200
4	190	280
5	250	370
6	320	430
7	370	460
8	420	470
9	450	470
10	460	470

5 Draw a graph to show the reaction progress for both the chips and the powder.

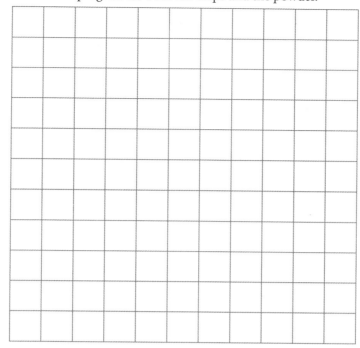

6 Compare the rate of each reaction.

_____

_____

_____

_____

7 a Identify the reaction that went faster.

_____

b Explain why that reaction went faster than the other one.

_____

_____

_____

_____

_____

8 a Predict the effect of using a higher concentration of acid on the reaction rate.

_____

_____

_____

b On your graph from question **5**, draw a dotted line to illustrate your answer to part **a** for the powder.

Explain how different factors affect reaction rate using collision theory.

Explain how molecular orientation and energy affect the success of a reaction.

Relate activation energy and temperature to the number of successful collisions.

1 Complete the table below by describing the effect of the action on the reaction rate and explaining why that occurs in terms of collision theory of reactions.

Action	Effect on reaction rate	Collision theory explanation
Heating the reactants		
Using a powdered solid rather than a lump		
Increasing the pressure on a gas reactant		
Cooling a liquid reactant		
Increasing the concentration of a liquid reactant		
Adding more water to dissolved reactants		
Adding a catalyst to dissolved reactants		
Increasing the volume of a gas reactant		
Dissolving a solid reactant first		

2 a When carbon monoxide and nitrogen dioxide molecules collide, they sometime react to form carbon dioxide and nitric oxide according to the equation:

$$CO(g) + NO_2(g) \rightarrow CO_2(g) + NO(g)$$

The reaction can be modelled using space-filling drawings of molecules, which are shown below in the correct orientation for a successful collision.

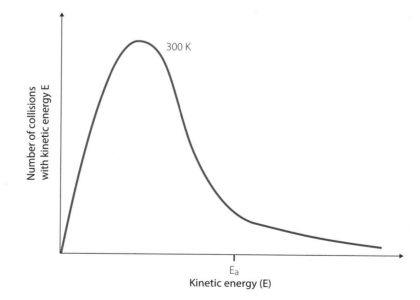

Successful collisions depend on the molecules having a combined kinetic energy equal to or greater than the activation energy for the reaction, and the collision occurring at the right orientation.

i Using the colour coding shown above, draw space-filling molecular model diagrams to illustrate the following types of collisions. Show the direction and speed of molecules by using arrows.

ii In the results column, draw the results of each of the collisions.

	Diagram	Results
Sufficient energy at the correct orientation		
Correct orientation but insufficient energy		
Sufficient energy but wrong orientation		
Insufficient energy and wrong orientation		

b The diagram below shows the distribution of kinetic energy for collisions between $CO(g)$ and $NO_2(g)$ at a temperature of 300 K. E_a is the activation energy for the reaction between these two molecules.

On the diagram, shade the area of the graph where the number of collisions has enough energy for a reaction.

c On the diagram, draw another graph showing the energy distribution for collisions if the temperature is increased to 500 K and shade (in a different colour) the areas under the graph where the number of collisions has enough energy for reaction.

300 K

Number of collisions with kinetic energy E

E_a

Kinetic energy (E)

WS 12.3 Analysing factors affecting reaction rate

LEARNING GOALS

Evaluate the use of experimental controls to collect valid and reliable data.

Assess risks.

Analyse data and assess error.

Identify relationships in data.

Draw conclusions based on an investigation.

Write balanced chemical equations.

For their Depth Study, a student decided to investigate factors that affect reaction rate. After researching the topic, they decided to conduct a series of experiments that looked at the effect of changing temperature and concentration. Because they had already done an investigation in class on changing surface area, they decided not to look at this factor; however, they used the same experimental set-up as they did in that investigation.

Their aim was to determine the effect of temperature and concentration on reaction rate. In order to see whether the effect is the same for different reactions, the experiments were repeated for two different chemical reactions.

Investigation	Experiment A	Experiment B
#1 HCl + Mg	Independent variable: Concentration of HCl mol L^{-1} (0.5, 1.0, 1.5, 2.0)	Independent variable: Temperature °C (25, 35, 45, 55, 65)
#2 HCl + CaCO$_3$	Independent variable: Concentration of HCl mol L^{-1} (0.5, 1.0, 1.5, 2.0)	Independent variable: Temperature °C (25, 35, 45, 55, 65)

For Investigation 1, they used the reaction between a strip of magnesium ribbon and hydrochloric acid. The rate was measured by timing how long it took to produce 30 mL of hydrogen gas. The two different experiments (A and B) were each conducted three times.

For Investigation 2, they used a reaction between hydrochloric acid and calcium carbonate. The rate was measured by timing how long it took to produce 30 mL of carbon dioxide gas. The two different experiments (A and B) were each conducted three times.

Their experimental set-up is shown below.

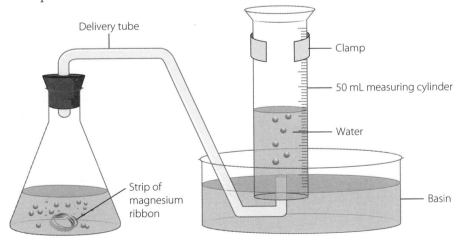

Delivery tube

Clamp

50 mL measuring cylinder

Water

Basin

Strip of magnesium ribbon

1 Explain what the student should do in the above experiments to ensure valid and reliable data.

_____

_____

_____

_____

_____

_____

_____

_____

_____

2 Complete the following risk assessment for the investigations.

What are the risks in doing this investigation?	How can you manage these risks to stay safe?

3 Write balanced chemical equations for each of the reactions used.

_____

_____

The results of the experiments for both investigations are given below.

Concentration factor

Concentration of HCl (mol L^{-1})	Time to produce 30 mL of gas (sec)							
	Investigation 1 A (HCl + Mg)				Investigation 2 A (HCl + CaCO$_3$)			
	Trial 1	Trial 2	Trial 3	Average	Trial 1	Trial 2	Trial 3	Average
0.50	306	302	305		360	365	363	
1.0	225	215	224		181	184	183	
1.5	63	61	62		118	100	116	
2.0	22	21	21		90	92	91	

Temperature factor

Temperature (°C)	Time to produce 30 mL of gas using 1.5 mol L⁻¹ HCl (sec)							
	Investigation 1 B (HCl + Mg)				Investigation 2 B (HCl + CaCO₃)			
	Trial 1	Trial 2	Trial 3	Average	Trial 1	Trial 2	Trial 3	Average
25	63	63	65		120	122	210	
35	32	31	32		65	65	66	
45	17	17	17		32	30	31	
55	9	9	10		18	17	18	
65	5	4	4		12	11	12	

4 a Calculate the averages for each set of results and write your values in the tables above.

   b Identify any results you did not include in your averages and explain why.

   _____

   _____

5 Describe the relationship between concentration and reaction rate.

   _____

   _____

6 a On the same piece of graph paper, draw two graphs for the results for temperature.

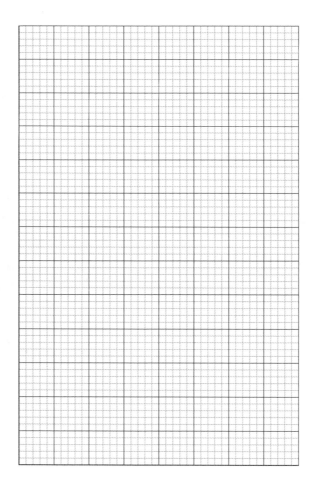

b Describe the relationship between temperature and reaction rate.

_____

_____

_____

7 a Identify what the student was trying to demonstrate.

_____

_____

_____

b Write a relevant conclusion based on their results.

_____

_____

_____

_____

_____

_____

_____

_____

_____

_____

8 Suggest which of the two reactions has the higher activation energy. Explain why.

_____

_____

_____

_____

_____

_____

Module three: Checking understanding

1 Match each term on the left with the appropriate description on the right by writing the number of the description next to the term.

A	Balanced equation		1	Occurs when atoms of two or more elements combine
B	Activation energy		2	Amount of solute dissolved in a given volume of solvent
C	Catalyst		3	Occurs when a compound breaks down into simpler chemicals
D	Collision theory		4	Energy of movement
E	Concentration		5	Explains reaction rates at a molecular level
F	Direct combination reaction		6	Reaction that occurs as written
G	Net ionic equation		7	Both sides have equal numbers of each type of atom
H	Combustion		8	How fast a chemical reaction is going
I	Decomposition reaction		9	How much one quantity changes with respect to another
J	Spectator ions		10	Occurs when an atom or molecule gains electrons
K	Precipitate		11	Measure of the average kinetic energy of particles
L	Kinetic energy		12	Occurs when an atom or molecule loses electrons
M	Neutralisation		13	Minimum energy required for a successful collision
N	Oxidation		14	The electrode at which reduction occurs
O	Spontaneous reaction		15	The reaction between an acid and a base
P	Salt bridge		16	The electrode at which oxidation occurs
Q	Reduction		17	Solid produced by a reaction between two clear solutions
R	Rate		18	Substance that in solution or in molten form conducts electricity
S	Electrolyte		19	Device that provides a path for the migration of ions in a galvanic cell
T	Reaction rate		20	Listing of the metals in order of increasing or decreasing reactivity
U	Surface area		21	Describes the oxidation and reduction processes separately in terms of electrons lost or gained
V	Temperature		22	Total area of all exposed surfaces of a solid reactant
W	Half-equation		23	Reaction in which oxygen reacts with a fuel to give an oxide
X	Activity series		24	Ions that do not participate in a reaction
Y	Anode		25	Substance that changes reaction rate without being consumed
Z	Cathode		26	Equation that excludes any spectator ions

2 From the following list of substances, choose one or more to answer the questions below. Each substance may be used more than once or not at all, and there may be more than one answer for each. Write a balanced chemical equation for each reaction using the chemicals identified.

calcium hydroxide (aq)	hydrochloric acid (aq)	zinc carbonate (s)	sulfur (s)	iron (s)
sodium carbonate (aq)	sodium nitrate (aq)	magnesium (s)	copper (s)	oxygen (g)

a Which would undergo a direct combination reaction when heated strongly together?

_____

_____

_____

9780170449564

b Which would undergo a decomposition reaction to produce a gas if heated strongly?

c Which would react together to produce hydrogen gas?

d Which would react together to produce a precipitate?

e Which would react to produce carbon dioxide gas and water?

f Which would react to produce a salt and water only?

3 When a clear solution of aluminium chloride is mixed with a clear solution of lead(II) nitrate, a white precipitate forms.

a Use solubility rules to predict the name and formula of the precipitate.

b Justify your prediction.

c Write a neutral species equation for the reaction.

d Write a complete ionic equation for the reaction.

e Write a net ionic equation for the reaction.

4 Students performed the following experiment to determine the relative reactivity of several metals. Pieces of each metal were added in turn to 5 mL of water and to 5 mL of 0.5 mol L^{-1} hydrochloric acid solution in test tubes. Each test tube was carefully observed to see whether any bubbles of colourless gas formed. Results are tabulated below. A tick means that a gas was formed; a cross means no gas appeared.

	Ca	Cu	Fe	Li	Ag	Zn
With water	✓	✗	✗	✓	✗	✗
With HCl	✓	✗	✓	✓	✗	✓

a Use the results in the table to arrange the metals into three groups – highly reactive, moderately reactive, not reactive. Justify your answer by writing balanced equations for any reactions that occur.

_____

_____

_____

_____

_____

_____

_____

_____

_____

b If there is more than one metal in each of the above categories, suggest an investigation you could do to distinguish between the reactivity of these metals.

_____

_____

_____

_____

c What conclusions can you draw from this information about the relative oxidising or reducing abilities of lithium, calcium, zinc, iron, copper and silver?

_____

_____

_____

_____

5 a Identify the species that is oxidised and the species that is reduced in the following reaction. Name the oxidant and reductant.

$$I_2O_5 + 3CO \rightarrow I_2 + 3CO_2$$

_____

_____

b Give the change in oxidation number for the oxidant and reductant.

_____

6 Use the following information to answer the questions below.

The half-equations represent the reduction reactions of three substances: A^{3+}, B^{2+}, C_2.

$$A^{3+}(aq) + e^- \rightarrow A^{2+}(aq)$$
$$B^{2+}(aq) + 2e^- \rightarrow B(s)$$
$$C_2(aq) + 2e^- \rightarrow 2C^-(aq)$$

Pairs of the six species, (A^{3+}, A^{2+}), (B^{2+}, B), (C_2, C^-), were mixed in test tubes and, if any observable reaction occurred, the results were noted. **Some** of these results are given in the table. The combinations given by I, II and III in the table were not investigated.

Oxidant ↓	Reductant →		
	$A^{2+}(aq)$	B(s)	$C^-(aq)$
$A^{3+}(aq)$	–	Yes	I
$B^{2+}(aq)$	II	–	III
$C_2(aq)$	X	Yes	–

Yes → observable reaction occurs
X → no observable reaction occurs
– → not tested

a Identify which of I, II and III would be expected to produce an observable reaction? Explain your reasoning.

_____

_____

_____

_____

_____

b List the order of strength of the oxidants, from weakest to strongest.

_____

7 a Draw a labelled diagram of the electrochemical cell made by combining a Mg, Mg^{2+} electrode with a Hg^{2+}, Hg electrode with a platinum wire as an inert electrode. Indicate which electrode is positive and show the direction of electron flow and migration of ions. Identify the anode and the cathode.

b Write equations for the half-reactions and the overall reaction that would occur in this cell.

_____

_____

_____

8 Suppose that you are given the following question:

'Determine the cell reaction and the voltage for the standard cell made by combining a Zn(s), Zn^{2+}(aq) with an Ag^+(aq), Ag(s)!

Explain what is wrong with the following attempted answer.

The relevant standard electrode potentials are:

$$Ag^+(aq) + e^- \rightarrow Ag(s) \qquad E° = +0.80 \text{ V}$$
$$Zn^{2+}(aq) + 2e^- \rightarrow Zn(s) \qquad E° = -0.76 \text{ V}$$

So, the Ag^+(aq), Ag(s) half-cell will compete most strongly for electrons and is the cathode.

So, the half-cell reactions can be represented by the following half-equations:

Cathode: $Ag^+(aq) + e^- \rightarrow Ag(s)$

Anode: $Zn(s) \rightarrow Zn^{2+}(aq) + 2e^-$

To obtain a balanced equation for the cell reaction, we need to multiply the Ag^+, Ag half-equation by two, so that the number of electrons taken up by Ag^+(aq) ions is the same as the number removed from Zn atoms. Then we obtain:

$$2Ag^+(aq) + Zn(s) \rightarrow 2Ag(s) + Zn^{2+}(aq)$$

And so: Cell voltage = $2 \times E°_{cathode} - E°_{anode} = 2 \times 0.80 \text{ V} - (-0.76 \text{ V}) = 2.36 \text{ V}$

_____

_____

_____

_____

_____

9 The graph below plots the rate at which hydrogen gas is formed when magnesium is reacted with dilute hydrochloric acid in the reaction:

$$Mg(s) + 2HCl(aq) \rightarrow MgCl_2(aq) + H_2(g)$$

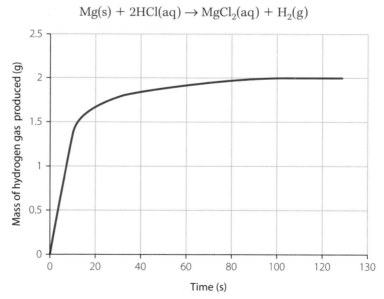

a How can you determine the reaction rate?

_____

_____

_____

b Describe the trend shown on the graph.

_____

_____

c Explain why that trend occurs.

_____

_____

_____

_____

10 A common method of preparing hydrogen for industrial purposes is to heat methane and steam with a catalyst at high temperature. The endothermic reaction is:

$$CH_4(g) + H_2O(g) \rightarrow 3H_2(g) + CO(g)$$

When 1 mole of methane was reacted with excess steam at 1100°C in a vessel of fixed volume, curve A in the following diagram was obtained. When the experiment was repeated at a different temperature, curve B was obtained.

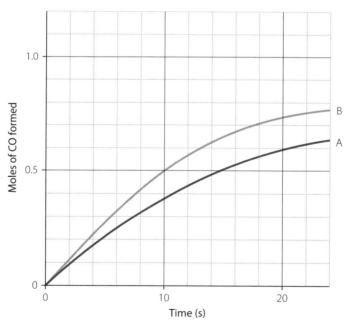

What do the initial slopes of the curves tell you about the initial rates at the two temperatures? Explain the reason for this difference.

_____

_____

_____

_____

_____

Reviewing prior knowledge

1 a Complete and balance the chemical reactions below.

    i _____ C_8H_{18} (l) + _____ O_2 (g) →_____ + _____

    ii $6CO_2$ (g) + $6H_2O$(l) → _____ + _____

    iii _____ NaOH(s) + _____ HCl(aq) →_____ + _____

  b For reaction **i**, identify which bonds are being broken, which bonds are being formed, and include the number of each of these bonds.

_____

_____

2 Explain how the rate of a reaction can be increased.

_____

_____

3 a Define the term 'catalyst'.

_____

_____

  b Explain how catalysts work.

_____

_____

  c This diagram shows the activation energy required for a chemical reaction. On the diagram, draw the profile for a catalysed path.

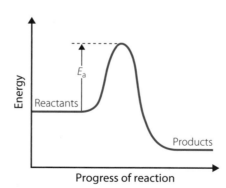

4 a Explain what is meant by a spontaneous reaction. Give an example of one reaction that is spontaneous and another that is not spontaneous.

_____

_____

_____

_____

b Describe how energy profile diagrams, such as the one in question **3c**, could be used to determine the spontaneity of a chemical reaction. Give an example to support your reason.

_____

_____

_____

5 a Name the type of bonds in the following substances.

i Within water molecules _____

ii Between water molecules _____

iii Within a NaCl crystal _____

b Explain why, in terms of bonding, an ionic substance dissolves in water.

_____

_____

_____

6 A student was asked to design an investigation to demonstrate the law of conservation of mass using a reaction between Mg and HCl.

Explain what the student needs to be aware of when using this reaction.

_____

_____

_____

_____

_____

7 a Explain the difference between the terms 'standard state of a substance' and 'standard conditions'.

_____

_____

_____

b What is the symbol used to show values that are taken for a substance in its standard state? Give an example of this use.

_____

_____

INQUIRY QUESTION: WHAT ENERGY CHANGES OCCUR IN CHEMICAL REACTIONS?

WS **13.1** **Investigating enthalpy changes**

STUDENT BOOK
Pages 319–30

LEARNING GOALS

Design valid calorimetry experiments with provided stimulus.

Calculate enthalpy changes using the heat capacity formula $q = mc\Delta T$.

Analyse and compare results with theoretical values to explain differences.

1 A student used calorimetry to determine the heat of solution of aluminium chloride ($AlCl_3$) and ammonium nitrate (NH_4NO_3) in 100 mL of water.

a Write a suitable method to determine the heat of solution of the aluminium chloride and ammonium nitrate using 100 mL of water.

_____

_____

_____

_____

_____

_____

_____

_____

_____

_____

_____

_____

_____

_____

_____

_____

_____

_____

_____

b The student absentmindedly recorded the results on a scrap piece of paper. Create a suitable table of their results.

Water – 100 mL

$AlCl_3$ – 10.0 g start 20.0°C finish 46.3°C

NH_4NO_3 – 4.9 g start 20.1°C finish 16.8°C

c For each solid, calculate the heat of solution in $kJ\,mol^{-1}$.

_____

_____

_____

_____

_____

_____

_____

_____

_____

_____

d Identify possible sources of error in this experiment.

_____

_____

_____

_____

e For each solid, write the chemical equation of dissolution, including the heat of solution, and identify each reaction as endothermic or exothermic.

_____

_____

_____

2 A student used calorimetry to determine the heat of combustion of three alkanols: methanol (CH_3OH), propanol (C_3H_7OH) and octanol ($C_8H_{17}OH$). The calorimeter consists of a small aluminium can filled with $200\,mL$ of distilled water, the temperature of which was taken before and after the experiment. The can was clamped to a retort stand above the flame of the spirit burner, whose weight is recorded before and after the experiment.

a Draw a labelled scientific diagram of the experimental set-up.

b Explain why an aluminium can was used rather than a polystyrene cup or glass container.

_____

_____

_____

c Write the balanced chemical equation for the complete combustion of the three alkanols used in this experiment.

_____

_____

_____

d Identify the following variables of this experiment.

   i Independent variable

   _____

  ii Dependent variable

   _____

 iii Three controls

   _____

   _____

   _____

   _____

e Construct a suitable Risk Assessment for this practical.

f The student recorded some of their results below.

Alkanol	Methanol (CH_3OH)	Propanol (C_3H_7OH)	Octanol ($C_8H_{17}OH$)
Mass of water (g)	200	200	200
Initial temperature of water (°C)	17.2	17.5	17.3
Final temperature of water (°C)	42.5	59.8	80.6
ΔT	+25.3	+42.3	+63.3
$q = mc\Delta T$ (J)			
Initial mass of spirit burner (g)	362.7	268.3	372.6
Final mass of spirit burner (g)	360.5	266.5	370.6
Mass of fuel burnt (g)			
ΔH (kJ mol^{-1})			

i Assuming the specific heat capacity of water is $4.18\,J\,°C^{-1}\,g^{-1}$, calculate the value for q when methanol was combusted. Calculate the molar heat of combustion of methanol (ΔH). Add these to the table above.

_____

_____

_____

_____

_____

_____

_____

_____

_____

ii Calculate the value for q and molar heat of combustion of propanol (ΔH) given the information in the table above. Add it to the table. If only 75% of the energy released from the combustion of propanol is transferred to the water and the rest was lost to surrounds, calculate the actual amount of energy released from the combustion of propanol.

_____

_____

_____

_____

_____

_____

_____

_____

iii Calculate the value for q and heat of combustion (ΔH) of octanol given the information in the table above. Add it to the table.

_____

_____

_____

_____

_____

_____

_____

iv The accepted theoretical value of octanol is $-5294.0\,\text{kJ mol}^{-1}$. Suggest two reasons why your calculated experimental value differs from the theoretical value.

_____

_____

_____

_____

_____

_____

_____

WS 13.2 Graphing the heat of combustion

1 A student conducted an experiment to determine the relationship between molar mass and the heat of combustion of different alcohols. The set-up of the experiment and the results are shown below.

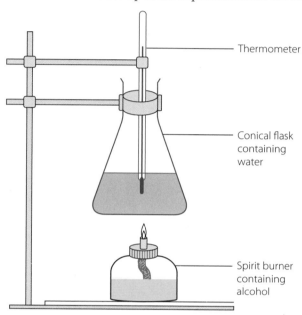

Thermometer

Conical flask containing water

Spirit burner containing alcohol

Alcohol	Molar mass (g mol⁻¹)	Experimental heat of combustion (kJ mol⁻¹)	Theoretical heat of combustion (kJ mol⁻¹)	Experimental heat of combustion (kJ g⁻¹)
Ethanol	46	160	720	
1-butanol	74	2300	2670	
1-heptanol	116	3880	4630	
1-octanol	130	4690	5285	

a On the grid below, graph both the experimental and theoretical heat of combustion (kJ mol^{-1}) against the molar mass of each alcohol.

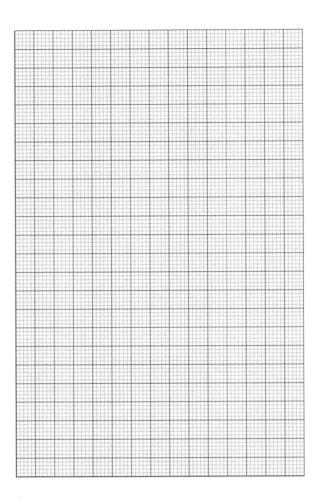

b Use your graph to estimate both the experimental and theoretical heat of combustion of 1-pentanol (C$_5$H$_{11}$OH).

_____

_____

> **HINT**
>
> Use a line of best fit to extrapolate data.

c Provide possible reasons for the differences between the experimental heat of combustion and the theoretical heat of combustion.

_____

_____

_____

_____

_____

_____

_____

d Suggest how this experiment could be improved to reduce experimental error.

_____

_____

_____

_____

e Complete the fifth column of the table, calculating the experimental heat of combustion in kilojoules per gram.

f It is often misleading to compare the energy release of fuels in per mole rather than in per gram. Explain why this is the case and identify where per gram information would be more useful.

_____

_____

_____

_____

2 Ethanol is being used in many countries as an alternative fuel to octane.

	Molar mass (g mol^{-1})	Theoretical heat of combustion (kJ mol^{-1})
Octane	114	5471

Given the data values in the table above and those from the graph in question **1a**, discuss the advantages and disadvantages of using ethanol as an alternative fuel source. Provide two suitable chemical equations.

_____

_____

_____

_____

_____

_____

_____

_____

_____

_____

_____

_____

_____

_____

_____

_____

LEARNING GOALS

Use an analogy to explain how a catalyst functions.

Model catalyst function in a reaction.

Explain the function of lactase in the human body from information within a model.

Model the function of an enzyme inhibitor.

1 You are at a party for a friend, but, aside from this friend, you don't know anyone else there. The night could naturally progress, with you bumping into a stranger – perhaps this will start a conversation or perhaps they will continue on. Alternatively, your friend could act as a matchmaker, introducing you to a stranger you have something in common with – you are guaranteed to start a conversation with this person and your friend is free to move on and mingle with others at the party.

Explain how this can be used as an analogy to describe the action of a catalyst.

_____

_____

_____

_____

_____

_____

_____

_____

_____

_____

2 In the Haber process, hydrogen gas reacts with nitrogen gas to produce ammonia gas according to the following equation:

$$3H_2(g) + N_2(g) \rightarrow 2NH_3(g)$$

This reaction proceeds too slowly to be economically viable, so an iron catalyst is added to the reaction vessel to provide a surface for the reaction to take place.

a Identify the iron catalyst as homogeneous or heterogeneous. Justify your choice.

_____

_____

b Draw a diagram to model the reaction process occurring with the iron catalyst in the Haber process.

3 Enzymes, such as lactase, are examples of biological catalysts. The function of lactase is modelled in the image below.

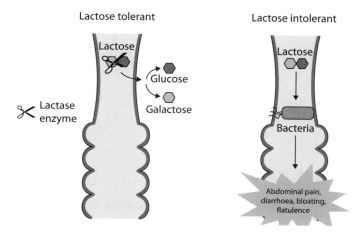

a With reference to the diagram above, describe the role and action of lactase in the human body.

_____

_____

_____

_____

_____

b The breakdown of lactose into glucose and galactose is a decomposition reaction that could naturally occur very slowly without the assistance of the lactase enzyme. Increasing the temperature of a chemical reaction can increase the rate of reaction. Explain whether or not increasing the body temperature would be a suitable alternative for sufferers of lactose intolerance.

_____

_____

_____

_____

_____

c Sketch an energy profile diagram for the decomposition of lactose with and without the presence of lactase.

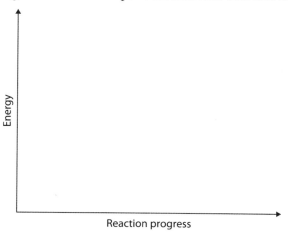

4 An enzyme inhibitor is a molecule that binds to an enzyme and decreases its ability to perform its function.

  a Draw a diagram to model this process.

  b How would the presence of an inhibitor affect the activation energy of a chemical reaction?

_____

_____

_____

14 Enthalpy and Hess's law

INQUIRY QUESTION: HOW MUCH ENERGY DOES IT TAKE TO BREAK BONDS, AND HOW MUCH IS RELEASED WHEN BONDS ARE FORMED?

WS 14.1 Understanding Hess's law

STUDENT BOOK
Pages 345–53

LEARNING GOAL

Calculate enthalpy of a reaction using Hess's law.

Hess's law states that for a chemical reaction the enthalpy change in going from reactants to products is constant, regardless of the particular set of reaction steps used to bring it about. Therefore, the enthalpy change is simply the difference in enthalpy between the products and the reactants and is independent on the path taken by the reaction.

1 Consider the reaction below.

$$W + X \rightarrow Y + Z \qquad \Delta H = ?$$

Its enthalpy cannot be measured directly. However, the enthalpy of the reactions below can be calculated:

i $W + X \rightarrow A$ $\qquad \Delta H_1 = b \, \text{kJ mol}^{-1}$

ii $A \rightarrow Y + Z$ $\qquad \Delta H_2 = c \, \text{kJ mol}^{-1}$

a Calculate the enthalpy, ΔH, for the reaction $W + X \rightarrow Y + Z$.

_____

_____

_____

b Complete the enthalpy diagram below to represent the reaction $W + X \rightarrow Y + Z$, using the information provided in the question.

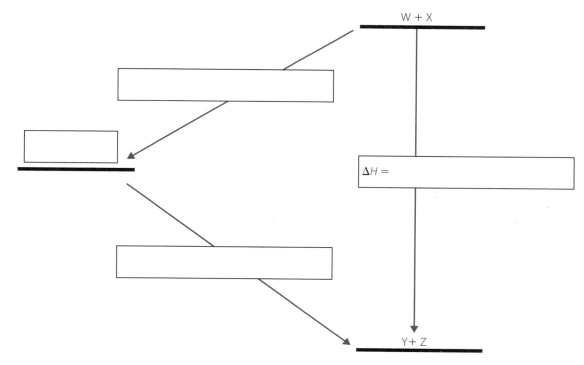

2 Calculate the enthalpy of formation of $NO_2(g)$ for the reaction shown, using the information provided in equations **i** and **ii**.

$$N_2(g) + O_2(g) \rightarrow 2NO_2(g) \quad \Delta H = ?$$

i $\quad \frac{1}{2}N_2 + \frac{1}{2}O_2 \rightarrow NO(g) \qquad\qquad \Delta H = +90\,kJ\,mol^{-1}$

ii $\quad 2NO_2(g) \rightarrow 2NO(g) + O_2(g) \qquad \Delta H = +112\,kJ\,mol^{-1}$

_____

_____

_____

_____

3 Calculate the enthalpy of combustion of ethane, C_2H_6, in $kJ\,mol^{-1}$, using the information provided.

$$2C_2H_6(g) + 7O_2(g) \rightarrow 4CO_2(g) + 6H_2O(l)$$

i $\quad C(s) + O_2(g) \rightarrow CO_2(g) \qquad\qquad \Delta H = +s\,kJ$

ii $\quad H_2(g) + \frac{1}{2}O_2(g) \rightarrow H_2O(l) \qquad \Delta H = +t\,kJ$

iii $\quad 2C(s) + 3H_2(g) \rightarrow C_2H_6(g) \qquad \Delta H = +u\,kJ$

_____

_____

_____

_____

_____

_____

4 Calculate the heat of formation of methane, CH_4, as shown in the equation below, using the equations **i** to **iii** provided.

$$C(s) + 2H_2(g) \rightarrow CH_4(g) \quad \Delta H = ?\,kJ$$

i $\quad C(s) + O_2(g) \rightarrow CO_2(g) \qquad\qquad\qquad \Delta H = +s\,kJ$

ii $\quad H_2(g) + \frac{1}{2}O_2(g) \rightarrow H_2O(l) \qquad\qquad \Delta H = +t\,kJ$

iii $\quad CH_4(g) + 2O_2(g) \rightarrow CO_2(g) + 2H_2O(l) \qquad \Delta H = +v\,kJ$

_____

_____

_____

_____

Interpret enthalpy diagrams for reactions.

Perform calculations to demonstrate the law of conservation of energy.

Use data to determine enthalpy of reaction.

Calculate percentage error in data, identify possible sources of error and suggest improvements.

1 The reactions and enthalpies for the formation of liquid water and steam are given.

$$2H_2(g) + O_2(g) \rightarrow 2H_2O(l) \qquad \Delta H_1 = -572\,kJ\,mol^{-1}$$
$$2H_2(g) + O_2(g) \rightarrow 2H_2O(g) \qquad \Delta H_2 = -484\,kJ\,mol^{-1}$$
$$H_2O(l) \rightarrow H_2O(g) \qquad \Delta H_3 = +44\,kJ\,mol^{-1}$$

Use the information provided to answer the questions below.

a The relationship between these reactions can be represented diagrammatically. Complete the missing substances in the following enthalpy diagram, showing the formation of one mole of water.

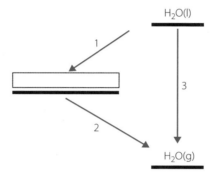

b Calculate the enthalpy for the paths labelled 1, 2 and 3 in the diagram above.

Path	Enthalpy (kJ mol⁻¹)
1	
2	
3	

c Use the data from the table in part **b**, to explain how this system demonstrates the law of conservation of energy in the above system.

_____

_____

_____

_____

_____

_____

_____

_____

2 A student conducted an investigation to attempt to demonstrate the law of conservation of energy. The student calculated the enthalpy values given using the data obtained in her investigation.

$$NaOH(s) \rightarrow NaOH(aq) \qquad\qquad \Delta H = -23.2\,kJ\,mol^{-1}$$

$$NaOH(s) + HCl(aq) \rightarrow NaCl(aq) + H_2O(l) \qquad \Delta H = -52.1\,kJ\,mol^{-1}$$

$$NaOH(aq) + HCl(aq) \rightarrow NaCl(aq) + H_2O(l) \qquad \Delta H = -28.9\,kJ\,mol^{-1}$$

a Draw an enthalpy diagram to represent the relationship between the reactions of sodium hydroxide in both solid and aqueous forms with hydrochloric acid.

b Explain whether the enthalpy values obtained by the student support the law of conservation of energy for the reaction of solid sodium hydroxide versus aqueous sodium hydroxide with hydrochloric acid. Refer to the data in your response.

c The theoretical value for the enthalpy of neutralisation of aqueous sodium hydroxide with hydrochloric acid is $-58.2\,kJ\,mol^{-1}$.

   i Calculate the percentage error in the student's result.

   ii Suggest possible reasons for the difference between experimental and theoretical results in this student's investigation.

   iii Suggest possible improvements that could be made to this investigation to minimise errors.

LEARNING GOALS

Identify equipment required to investigate heat of neutralisation.

Write an equation for a neutralisation reaction.

Draw an enthalpy diagram for a neutralisation reaction.

Use data from neutralisation experiments to calculate $q = mc\Delta T$ and ΔH.

Perform calculations to demonstrate the law of conservation of energy.

Determine if the heat of neutralisation between two aqueous solutions of a strong acid and strong base is the same as when the strong base is in a solid form.

A student performed two separate experiments to test the law of conservation of energy, which states that the energy change in going from reactants to products is the same regardless of the path taken. In the first experiment, the student reacted two aqueous solutions of hydrochloric acid and potassium hydroxide, while in the second experiment solid potassium hydroxide was added to hydrochloric acid.

The details of the experiments conducted are as follows.

Experiment 1

A 54 mL solution of 0.200 mol L^{-1} HCl was reacted with 54 mL of 0.200 mol L^{-1} KOH. The temperature changed from 20.00°C to 21.36°C.

Experiment 2

A 54 mL solution of 0.200 mol L^{-1} HCl was reacted with 0.6060 g of KOH. The temperature changed from 20.00°C to 23.72°C.

1 List the equipment the student would need for each experiment.

Experiment 1	Experiment 2

2 Write a balanced equation for each experiment.

a Experiment 1

b Experiment 2

3 Draw an enthalpy diagram for the reaction of potassium hydroxide with hydrochloric acid.

4 Assuming heat capacity of each solution is $4.18\,JK^{-1}g^{-1}$, calculate the heat of reaction for each experiment.

   a Experiment 1

_____

_____

_____

_____

_____

_____

   b Experiment 2

_____

_____

_____

_____

_____

5 Explain whether experiments 1 and 2 demonstrate the law of conservation of energy.

_____

_____

_____

_____

_____

_____

_____

_____

_____

_____

6 The student performed a third experiment in which she reacted 54 mL of 0.200 mol L^{-1} HCl with 54 mL of 0.100 mol L^{-1} KOH.

a Explain what effect changing the concentration of the potassium hydroxide solution will have on the molar heat of reaction. Include an equation in your answer.

_____

_____

_____

_____

_____

_____

b When the student performed the reaction, the temperature changed to 20.68°C from 20.00°C. Calculate the molar heat of reaction. Does your answer support your response to part a above?

_____

_____

_____

_____

_____

_____

_____

_____

_____

_____

WS 14.4 Solving problems for Hess's law

Draw enthalpy cycle diagrams for reactions using modelling.

Calculate enthalpy of reaction from enthalpy cycle diagrams to determine whether reaction is exothermic or endothermic.

Identify exothermic or endothermic reactions from energy diagrams.

Refer to bond enthalpy to determine whether a reaction is exothermic or endothermic.

Investigate Hess's law in quantifying the enthalpy change for a stepped reaction using standard enthalpy change data and bond energy data.

Perform Hess's law calculations.

1   A chemical reaction, R_1, can be modelled using the following shapes. The enthalpy, ΔH_1, of the reaction, R_1, shown, cannot be calculated directly.

Reaction, R_1

The enthalpy of the reactions below are shown.

$$\blacksquare \quad + \quad \blacktriangle \quad \longrightarrow \quad \text{⬡} \qquad \Delta H_2 = 50 \text{ kJ mol}^{-1}$$

$$\text{⬡} \quad \longrightarrow \quad \bullet \quad + \quad \text{⬠} \qquad \Delta H_3 = -110 \text{ kJ mol}^{-1}$$

a   Draw an enthalpy cycle for the reaction shown below, including ΔH_1, ΔH_2 and ΔH_3.

$$\blacksquare \quad + \quad \blacktriangle \quad \longrightarrow \quad \text{⬠} \quad + \quad \bullet \qquad \Delta H_1$$

b   Using the information above, calculate the enthalpy, ΔH_1, for the reaction R_1.

_____

_____

_____

_____

c   Is the reaction, R_1, endothermic or exothermic? Explain.

_____

_____

d Draw a graphical representation of enthalpy changes for reaction, R_1. In your diagram show the relationship between enthalpies of reactants and products and ΔH for the reaction.

2 The energy diagrams below, labelled A and B, show the enthalpy changes that occur during two chemical reactions.

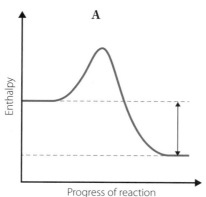

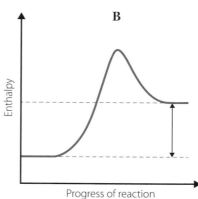

a Match the two following reactions to the graphs above.

   i $N_2(g) + O_2(g) \rightarrow 2NO(g)$ ΔH = positive value

   ii $N_2(g) + 3H_2(g) \rightarrow 2NH_3(g)$ ΔH = negative value

b On the appropriate energy profile diagram, label the energy of reactants and products respectively for each reaction.

c Identify and label the change in enthalpy, ΔH, for each graph.

d For each reaction in parts **ai** and **aii** above, compare the energy required to break and form bonds with respect to the reactants and products.

e For each reaction in parts **ai** and **aii** above, calculate the enthalpy of the reaction as *written*, from the table of standard enthalpies of formation given below.

i $N_2(g) + O_2(g) \rightarrow 2NO(g)$

ii $N_2(g) + 3H_2(g) \rightarrow 2NH_3(g)$

Table 14.1 Some standard enthalpies of formation at 298 K

Substance	$\Delta H^\ominus$ (kJ mol⁻¹)	Substance	$\Delta H^\ominus$ (kJ mol⁻¹)	Substance	$\Delta H^\ominus$ (kJ mol⁻¹)
$Al_2O_3(s)$	−1670	$HI(g)$	+26	$N_2O_4(g)$	+9
$BaSO_4(s)$	−1465	$H_2(g)$	0	$NH_3(g)$	−46
$Br_2(g)$	+31	$H_2O(g)$	−242	$NH_4Cl(s)$	−314
$CaO(s)$	−636	$H_2O(l)$	−285	$NaCl(s)$	−411
$Ca(OH)_2(s)$	−987	$H_2O_2(g)$	−136	$NaOH(s)$	−425
$CaCO_3(s)$	−1207	$H_2O_2(l)$	−188	$Na_2SO_4(s)$	−1385
$CaCl_2(s)$	−795	$H_2O_2(aq)$	−191	$O_2(g)$	0
$CO(g)$	−111	$H_2S(g)$	−21	$O_3(g)$	+143
$CO_2(g)$	−393	$H_2SO_4(l)$	−814	$PBr_3(l)$	−185
$Fe_2O_3(s)$	−823	$I_2(g)$	+62	$Si(s)$	0
$HBr(g)$	−36	$I_2(aq)$	+23	$SiCl_4(l)$	−687
$HCl(g)$	−92	$NO(g)$	+90	$SO_2(g)$	−297
$HF(g)$	−271	$NO_2(g)$	+33	$SO_3(g)$	−396

9780170449564

15 Entropy and Gibbs free energy

INQUIRY QUESTION: HOW CAN ENTHALPY AND ENTROPY BE USED TO EXPLAIN REACTION SPONTANEITY?

WS 15.1 Modelling and interpreting entropy

STUDENT BOOK
Pages 365–75

LEARNING GOALS

Compare entropy between substances.

Explain entropy changes in systems.

Assess representations designed to show entropy changes.

Identify a relationship between entropy and temperature.

Identify a relationship between entropy and molecular structure.

1 For each of the following pairs of substances, identify which substance would have the higher entropy. Give reasons for your decisions.

   a solid wax or molten wax

   _____

   _____

   b $Br_2(l)$ or $Br_2(g)$

   _____

   _____

   c pentane $C_5H_{12}(l)$ or octane $C_8H_{18}(l)$

   _____

   _____

   d silver(s) and silver chloride(s)

   _____

   _____

2 For each of the following reactions, explain whether there would be an increase or a decrease in entropy.

   a $2SO_2(g) + O_2(g) \rightarrow 2SO_3(g)$

   _____

   _____

   b $2Mg(s) + O_2(g) \rightarrow 2MgO(s)$

   _____

   _____

   c $CaCO_3(s) \rightarrow CaO(s) + CO_2(g)$

   _____

   _____

3 Students were asked to produce representations (models, diagrams or photographs) that demonstrated an increase in entropy of a system.

Student A presented the following diagram showing diffusion of a monatomic gas.

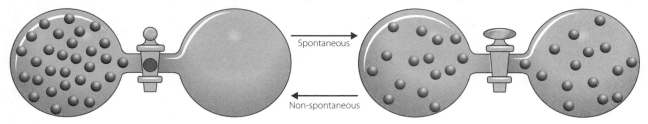

Student B presented the following diagram showing the dissolving of sodium chloride in water.

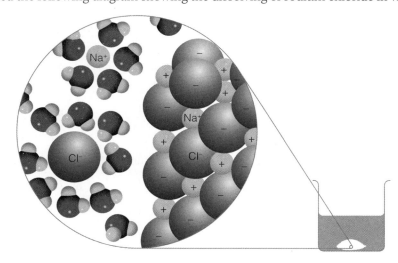

Student C presented the following drawings, before and after reaction, showing the reaction between two solids, barium hydroxide and ammonium chloride, which was demonstrated by their teacher.

Before reaction   After reaction

a Assess each student's submission, explaining how their presentation demonstrated an increase in entropy.

_____

_____

_____

_____

_____

_____

_____

_____

_____

_____

b For each submission above, suggest what other information could have been provided to demonstrate entropy is a/the driver of each process shown.

_____

_____

_____

_____

_____

_____

_____

_____

_____

4 The graph below shows generally how the entropy of a single pure substance changes as the substance is heated and undergoes successive changes of state.

a Draw particle diagrams to show how entropy of a system increases as a substance changes from a solid to liquid to gas.

b At what temperature would it be possible for the entropy of a substance to be zero?

_____

c Suggest a reason why the entropy of a particular state increases with temperature.

_____

_____

_____

d Use researched data and the graph above to support or refute the statement:

'The difference in entropy for any two gases is much smaller than the difference between the entropies for any substance as a liquid and as a gas.'

_____

_____

_____

_____

_____

5 a Diamond and graphite are both allotropes of carbon with the structures shown below.

The structure of diamond

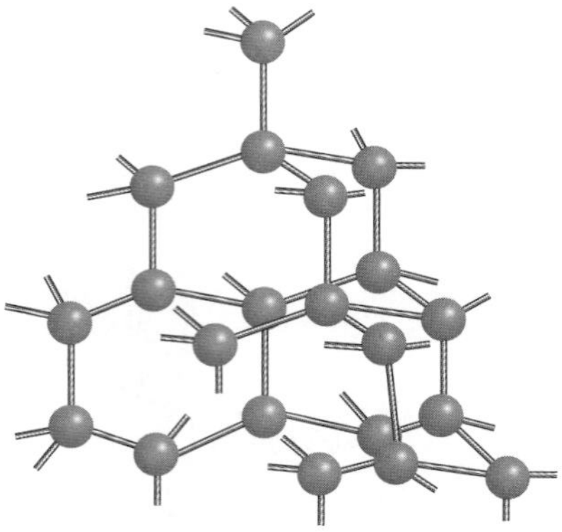

The structure of graphite

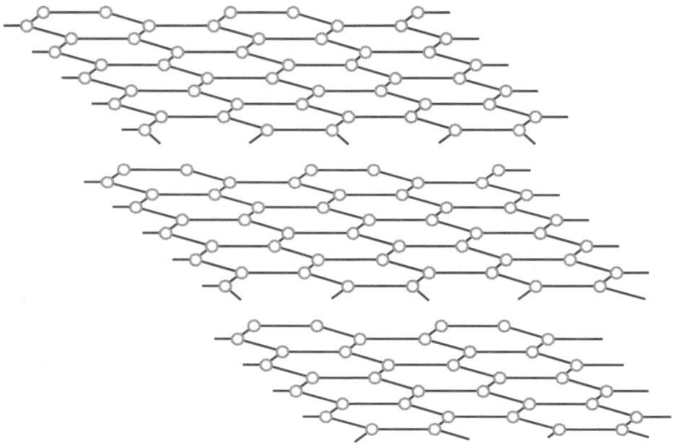

Use your knowledge of these structures to explain why the entropy of diamond ($S^\ominus = 2.4 \, J \, K^{-1} \, mol^{-1}$) is less than that of graphite. ($S^\ominus = 5.7 \, J \, K^{-1} \, mol^{-1}$)

_____

_____

_____

_____

_____

b

Compound	Methane (g) CH$_4$	Ethane (g) C$_2$H$_6$	Propane (g) C$_3$H$_8$	Butane (g) C$_4$H$_{10}$
Standard entropy (S$^\ominus$) J K^{-1}mol^{-1}	186	229.5	270	310
Compound	Methanol (l) (CH$_3$OH)	Ethanol (l) (CH$_3$CH$_2$OH)	Propanol (l) (CH$_3$CH$_2$CH$_2$OH)	Butanol (l) (CH$_3$CH$_2$CH$_2$CH$_2$OH)
Standard entropy (S$^\ominus$) J K^{-1}mol^{-1}	127	160	197	228

Use the data provided in the table above to complete the following.

i Write a generalisation between entropy and the complexity of molecular structures.

ii Write a generalisation between entropy and molecules with hydrogen bonding. (The alcohols, methanol, ethanol, propanol and butanol, listed above exhibit hydrogen bonding between molecules.)

LEARNING GOALS

Evaluate enthalpy and entropy changes in reactions.

Apply data to predict reaction direction.

Predict reaction spontaneity using Gibbs free energy.

Assess validity of a given response.

Predict the effect of temperature changes on a reaction.

1 Use data to evaluate the statement:

'Evaporation of water occurs spontaneously because the entropy drive for evaporating is much larger than the energy drive for liquefying.'

	$H_2O(g)$	$H_2O(l)$
$\Delta H_f^{\ominus}$ (kJ mol^{-1})	−242	−286
$S^{\ominus}$ (J K^{-1} mol^{-1})	189	70

_____

_____

_____

_____

_____

_____

_____

_____

_____

_____

_____

2 For each of the following reactions:

    A $H_2O_2(l) \rightarrow H_2O(l) + \frac{1}{2}O_2(g)$ exothermic

    B $N_2(g) + O_2(g) \rightarrow 2NO_2(g)$ endothermic

    C $N_2O(g) + 2H_2O(l) \rightarrow NH_4NO_3(s)$ endothermic

    D $2Cu(s) + S(s) \rightarrow Cu_2S(s)$ exothermic

    E $HBr(g) \rightarrow \frac{1}{2}H_2(g) + Br(l)$ endothermic

a identify the direction of the energy drive

b identify the direction of the entropy drive

c predict whether the reaction will proceed as written, giving a reason for your prediction.

Use the table on the next page to record your answers.

Reaction	Energy drive direction	Entropy drive direction	Prediction
A			
B			
C			
D			
E			

3 a Use the following data for standard conditions for the reactions A–E in question **2** above to check your predictions. If you were unable to predict, use the data to adjust your prediction.

Reaction	$\Delta H^{\ominus}$ (kJ mol^{-1})	$\Delta S^{\ominus}$ (JK^{-1}mol^{-1})
A	−98	+63
B	+180	+25
C	+124	−209
D	−80	+121
E	+36	−199

b Which predictions, if any, did you change and why?

c Comment on the spontaneity of each of the above reactions.

_____

_____

_____

4 A student was given the following question and associated data.

(I) Calculate $\Delta G^\ominus$ for the reaction:

$$2K^+(aq) + CO_3^{2-}(aq) \rightarrow K_2CO_3(s)$$

(II) Comment on the solubility of this salt in water.

Substance	$\Delta H^\ominus$ (kJ mol^{-1})	$S^\ominus$ (JK^{-1}mol^{-1})
K^+(aq)	−252	101
CO_3^{2-}(aq)	−675	−50
K_2CO_3(s)	−1151	156

The student's answer is given below:

(I) $\Delta G^\ominus = \Delta H^\ominus - T\Delta S^\ominus = -224 - (298 \times 0.105) = -150$ kJmol^{-1}

(II) The precipitate will form spontaneously and is not very soluble in water.

However, according to solubility tables, all potassium salts are soluble.

What should the correct answer be and what mistake/s did the student make?

_____

_____

_____

_____

_____

_____

_____

5 The story of Napoleon's buttons demonstrates the importance of understanding chemistry. The buttons fell apart on the coats of members of Napoleon's army invading Russia in 1812. The buttons were made of tin. Tin exists as two allotropes. Sn (silver) is a metallic form while Sn (grey) is much more brittle. When subjected to cold conditions, the surface of silvery metallic tin can become crumbly and fall apart as it changes to the grey allotrope. This is often referred to as tin disease.

$$Sn(silver) \rightarrow Sn(grey)$$

Substance	$\Delta H^\ominus$ (kJ mol^{-1})	$S^\ominus$ (JK^{-1}mol^{-1})
Sn (grey)	−2	44
Sn (silver)	0	51

a Use the data to calculate $\Delta G^\ominus$ at:

   i 22°C

   ii −30°C.

_____

_____

_____

_____

_____

_____

b Comment on the effect of temperature on the change from silver to grey allotrope.

_____

_____

c Explain any assumptions that have been made in calculating $\Delta G^\ominus$.

_____

_____

_____

6 Under appropriate conditions, calcium carbonate decomposes to calcium oxide according to the following reaction:

$$CaCO_3(s) \rightarrow CaO(s) + CO_2(g)$$

The enthalpy and entropy values for the reactants and products are given in the table below.

Substance	$\Delta H^\ominus$ (kJ mol^{-1})	$S^\ominus$ (J K^{-1} mol^{-1})
$CaCO_3(s)$	−1207	93
$CaO(s)$	−635	38
$CO_2(g)$	−394	214

a Will calcium carbonate decompose under normal laboratory conditions? Use data to support your answer.

_____

_____

_____

_____

_____

_____

b At what temperature will calcium carbonate decompose?

_____

_____

_____

_____

> **HINT**
>
> For this reaction to occur, $\Delta G^\ominus$ must be negative (i.e. less than 0).

Predict entropy changes using balanced chemical equations and standard reference values.

Solve problems related to entropy and Gibbs free energy using standard reference values.

Write balanced chemical equations.

Identify relationships between entropy and compound properties and bonding.

1 The reaction below occurs in the direction written:

$$MgCl_2(aq) + 2NaOH(aq) \rightarrow Mg(OH)_2(s) + 2NaCl(aq) \qquad Endothermic$$

a Explain how this information can be used to decide on the direction and size of the entropy of this reaction.

b **i** Write the net ionic equation for this reaction.

ii Given the following data, calculate $\Delta S^{\ominus}$ for the reaction in part **bi**.

Substance	$S^{\ominus}$ ($J\,K^{-1}\,mol^{-1}$)
$Mg^{2+}(aq)$	−137
$OH^-(aq)$	−11
$Mg(OH)_2(s)$	63

c Justify why the spectator ions can be ignored in the calculation of entropy.

d Use the data from part **bii** and that provided below to determine if there is a relationship between entropy of a precipitate of the same cation and its solubility.

For $MgCO_3(s)$ $S^{\ominus} = 66\,J\,K^{-1}\,mol^{-1}$ Solubility $= 0.06\,g/100\,g$

For $MgF(s)$ $S^{\ominus} = 57\,J\,K^{-1}\,mol^{-1}$ Solubility $= 0.013\,g/100\,g$

For $Mg(OH)_2(s)$ Solubility $= 0.0012\,g/100\,g$

2 a Consider the following statement:

'The reason many cations and anions react to form precipitates is because of the disorder that results when the cations and anions release the water molecules they are bonded to in their aqueous states.'

$$e.g.\ Mg(H_2O)_{36}^{2+} + CO_3(H_2O)_{28}^{2-} \rightarrow MgCO_3(s) + 64H_2O$$

Use your understanding of the bonding between ions and water to explain what this means. You may wish to include a diagram in your explanation.

----

b Use the information above to explain why $S° = -137\ J\,K^{-1}mol^{-1}$ for Mg^{2+} ion.

----

3 The reaction between sodium hydrogen carbonate (bicarb soda) and citric acid occurs spontaneously despite the fact that $\Delta H° = +80\ kJ\,mol^{-1}$. Propose an explanation as to why this reaction occurs spontaneously.

----

4 a A teacher demonstrated the reaction between barium hydroxide and ammonium chloride to show an example of a spontaneous endothermic reaction.

The equation for the reaction is:

$$Ba(OH)_2.8H_2O(s) + 2NH_4Cl(s) \rightarrow 2NH_3(g) + 10H_2O(l) + BaCl_2(s)$$

Use the data below to calculate the value of $\Delta G^\ominus$.

Substance	$\Delta H^\ominus$ (kJ mol^{-1})	$S^\ominus$ (J mol^{-1} K^{-1})
$Ba(OH)_2.8H_2O(s)$	−3345	427
$NH_4Cl(s)$	−314	95
$BaCl_2(s)$	−895	124
$NH_3(g)$	−46	192
$H_2O(l)$	−286	70

_____

_____

_____

_____

_____

_____

_____

_____

_____

_____

b Predict the effect of lowering temperature on the spontaneity of this reaction.

_____

_____

_____

_____

_____

Module four: Checking understanding

1 Match the terms on the left with the most appropriate statement on the right by writing the number of the statement in the box next to the term.

A	Conservation of energy		1	Measurement and calculation of heat changes
B	Endothermic reaction		2	Reaction that releases heat to the environment
C	Change in enthalpy		3	The study of heat, energy and motion
D	Exothermic reaction		4	Heat absorbed when one mole of a substance dissolves in water
E	Heat of combustion		5	Randomness of a system
F	Specific heat capacity		6	Energy cannot be created or destroyed, only transformed
G	Spontaneous reaction		7	Heat released when one mole of a substance undergoes complete combustion
H	Thermochemical equation		8	Equation that includes both chemicals and the energy change
I	Heat of solution		9	The energy change for a particular reaction is independent of the reaction steps
J	Calorimetry		10	Heat absorbed per mole of an identified substance in a reaction at a constant pressure
K	Bond energy		11	Heat needed to raise the temperature of 1 gram by 1 K
L	Hess's law		12	Measure of the net result of energy and entropy drives for a reaction
M	Stability		13	Reaction that occurs at room temperature without any continual heating
N	Entropy		14	Energy required to break a particular chemical bond
O	Gibbs free energy		15	Measure of how difficult It is to decompose a compound
P	Thermodynamics		16	Reaction in which the energy of the products is less than that of the reactants

2 Predict the direction of greatest entropy for the following reactions. Give a supporting reason for your prediction.

a $CuSO_4.5H_2O(s) \rightarrow CuSO_4(s) + 5H_2O(l)$

_____

_____

b $HCl(g) + NH_3(g) \rightarrow NH_4Cl(s)$

_____

_____

c $2SO_3(g) \rightarrow 2SO_2(g) + O_2(g)$

_____

_____

d $H_2(g) + I_2(g) \rightarrow 2HI(g)$

_____

_____

3 The heat of combustion of hexane, C_6H_{14}, was measured as follows.

A small spirit burner (glass container with a cotton wick) containing hexane had a mass of 164.2 g. It was placed under an aluminium can of mass 58.3 g containing 700 mL water at 14.7°C. The spirit burner was lit and used to heat the can and the water it contained. When the temperature of the can and water was 36.3°C, the burner was extinguished. The burner with the remaining hexane was then weighed; the final mass was 162.3 g.

Specific heat capacities of aluminium and water are 0.90 and 4.18 $J\,K^{-1}\,g^{-1}$ respectively.

a Write a balanced equation for the combustion of hexane.

_____

b Calculate the heat of combustion of hexane, both per gram and per mole.

_____
_____
_____
_____
_____
_____
_____
_____

c Use the information below to calculate the theoretical molar heat of combustion of hexane.

The structure of hexane is:

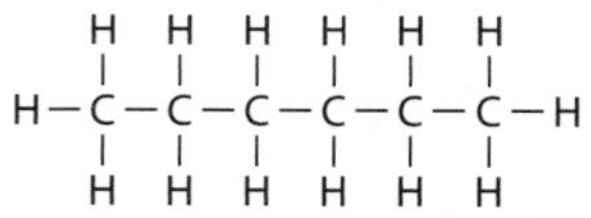

Bond energies (in kJ mol⁻¹) at 298.2 K					
C–C	348	C–H	413	O–H	463
		C=O	805	O=O	498

_____
_____
_____
_____
_____
_____
_____
_____
_____
_____
_____

d According to data tables, the molar heat of combustion of hexane is $-4163 \text{ kJ mol}^{-1}$. Compare the values obtained in parts **b** and **c** with the value found in the data table and suggest reasons for any differences.

HINT

The values used for bond energies are averages and not specific to particular compounds.

_____

_____

_____

_____

_____

_____

_____

4 a Find the ΔH for the reaction below, given the following reactions and subsequent ΔH values:

$$N_2H_4(l) + H_2(g) \rightarrow 2NH_3(g)$$

$$N_2H_4(l) + CH_4O(l) \rightarrow CH_2O(g) + N_2(g) + 3H_2(g) \qquad \Delta H_1 = -37 \text{ kJ}$$

$$N_2(g) + 3H_2(g) \rightarrow 2NH_3(g) \qquad \Delta H_2 = -46 \text{ kJ}$$

$$CH_4O(l) \rightarrow CH_2O(g) + H_2(g) \qquad \Delta H_3 = -65 \text{ kJ}$$

_____

_____

_____

_____

_____

b i Calculate the standard free-energy change ($\Delta G^\ominus$) at 25°C for the reaction above, given the entropies of the products and reactants are:

$$S^\ominus(N_2H_4) = 121.2 \text{ J K}^{-1} \text{mol}^{-1}$$

$$S^\ominus(NH_3) = 193 \text{ J K}^{-1} \text{mol}^{-1}$$

$$S^\ominus(H_2) = 130.7 \text{ J K}^{-1} \text{mol}^{-1}.$$

_____

_____

_____

_____

_____

_____

ii Is the reaction spontaneous at this temperature?

_____

5 The reaction between aqueous silver nitrate and potassium chloride is given below:

$$AgNO_3(aq) + KCl(aq) \rightarrow AgCl(s) + KNO_3(aq)$$

You have been asked to determine the entropy of AgCl, but that information is not listed in your tables. However, you have been able to obtain the following information.

	AgNO₃	KCl	AgCl	KNO₃
$\Delta H^{\ominus}_f$ (kJ mol⁻¹)	−124	−437	−127	−495
$S^{\ominus}$ (J K⁻¹mol⁻¹)	141	83		133
ΔG_f (kJ mol⁻¹)	−33	−409	−110	−395

What is $S^{\ominus}$ for AgCl?

_____

_____

_____

_____

_____

_____

_____

_____

_____

_____

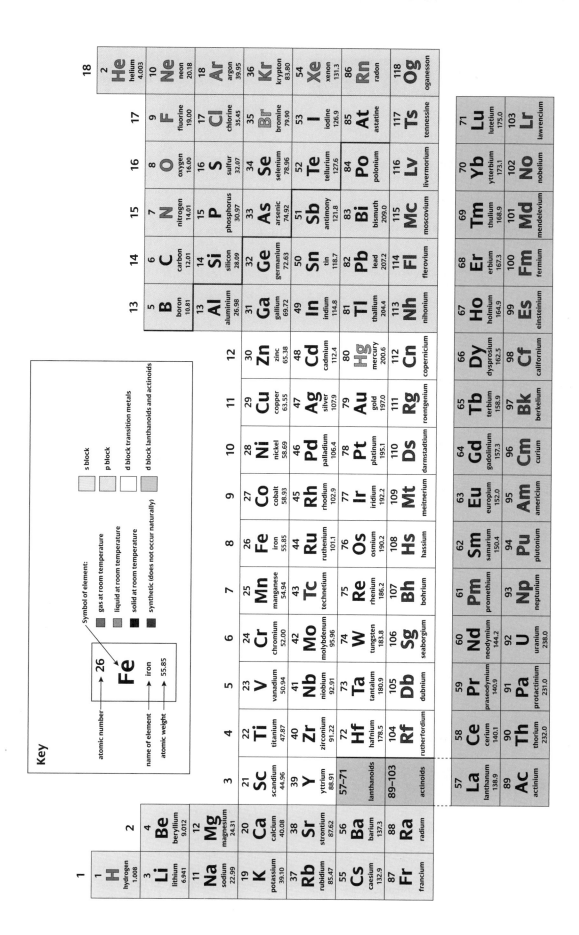

ANSWERS

Fully worked solutions are provided below to demonstrate the steps necessary to reach the required answer. Worked solutions help you independently review your own answers.

An introduction to working scientifically and depth studies

WS 1.1 PAGE 1

1. a How does changing the concentration of the $4\,mol\,L^{-1}$ nitric acid by diluting it to $2\,mol\,L^{-1}$, $1\,mol\,L^{-1}$ and $0.5\,mol\,L^{-1}$ affect the rate of reaction with $2\,g$ of marble chips?
   b If the $4\,mol\,L^{-1}$ nitric acid is diluted, then the rate of reaction with $2\,g$ of marble chips will decrease.
   c The concentration of the acid
   d The time taken for the same volume of gas to be produced for each concentration
   OR
   The volume of gas produced in a set amount of time
   e Volume of nitric acid
   Mass of marble chips
   Temperature of nitric acid
   Surface area of marble chips
   Same equipment

2. a How does the temperature of the solution affect the time taken for the cross to disappear?
   b If the temperature of the solution is increased, then the time taken for the cross to disappear will decrease.
   c The temperature of the system
   d The time taken for the cross to disappear/for the solution to become opaque
   e Volume and concentration of sodium thiosulfate
   Volume and concentration of hydrochloric acid
   Same equipment
   Same size black cross

3. a How does the conductivity of each dissolved solid in water compare?
   b If the solid is ionic, then when dissolved in water the solution will conduct electricity.
   OR
   If the solid is covalent molecular, then when dissolved in water the solution will not conduct electricity.

4. a What colour flame will sample X produce?
   b Any of the following answers is acceptable.
   If the solid X produces a dull red colour flame, then it contains lithium.
   If the solid X produces a yellow colour flame, then it contains sodium.
   If the solid X produces a lilac colour flame, then it contains potassium.
   If the solid X produces a brick red colour flame, then it contains calcium.
   If the solid X produces a scarlet colour flame, then it contains strontium.
   If the solid X produces an apple green colour flame, then it contains barium.
   If the solid X produces a blue-green colour flame, then it contains copper.
   If the solid produces a particular colour flame, then it can be identified.

5. a What is the effect on boiling point when salt is added to distilled water?

   b If $10\,g$ of salt is added to $100\,mL$ of distilled water, then the boiling point of the water will be higher than that of pure distilled water.
   c The amount of salt added
   d The boiling point of water
   e Volume of water
   Same size beaker
   Same thermometer range
   Same depth of immersion of thermometer bulb

WS 1.2 PAGE 4

1.

a	A student read the volume in a measuring cylinder from above instead of at eye level.	Mistake
b	The top-loading balance was not zeroed before measuring and recording multiple masses.	Systematic error
c	A student recorded the titre volumes in an experiment as 24.11 mL, 24.15 mL, 24.14 mL and 36.28 mL. What type of error may have contributed to the value of 36.28 mL?	Mistake/random error
d	Repetition of an experiment does not minimise this type of error.	Systematic error
e	Repetition of an experiment does minimise this type of error.	Random error

2. a Group 1 because the mass of $550\,g$ is an outlier.
   b Group 2 because the mass measurements were close to each other but not close to the actual value of $122\,g$.
   c Group 3 because their mass measurements were close to each other and matched the actual value of $122\,g$.
   d Group 2 because all their values were $10\,g$ more than the true value. This could have been due to faulty calibration of the balance or not taring the container used for weighing.
   e Group 3 because their data was replicated and matched the actual value.

3. The % difference $= \dfrac{|2.0 - 2.5|}{2.5} \times 100 = 20\%$

4. a Y   b Z   c W   d X

5. a The accuracy of the conductivity meter is more important because the actual or true conductance of each solution needs to be observed.
   b Title: The conductivity of water with varying amounts of salt
   x axis: Mass of salt added (g)
   y axis: Conductance (S/m)
   c The experiment is not reliable as only one data point appears to be plotted for each mass of salt. The results are not accurate because the conductivity for the various amounts of salt do not match the theoretical values. The experiment is somewhat valid because the general trend line is similar to the theoretical values.
   d The difference could be due to systematic error because there is a consistent difference between the theoretical and experimental data, which indicates other variables were controlled.

1 a ±0.10

  b This means that the true value is anywhere between 8.95 mL and 8.75 mL.

2 a It is a piece of analogue equipment.

  b **i** When reading off the 2.0 V scale, the uncertainty is half of the smallest division, 0.05 V, which is ± 0.025 V.

    ii The uncertainty is half the smallest division, which is half of 0.5 V and it is ±0.25 V.

3 a The top-loading balance is a piece of digital equipment.

  b It measures mass to two decimal places. The smallest scale division is 0.01 g; therefore, the uncertainty is ± 0.01 g.

4 a A: Uncertainty = 0.01; Mass = 199.92;
Mass range = 199.91−199.93
B: Uncertainty = 0.01; Mass = 198.74;
Mass range = 198.73−198.75
C: Uncertainty = 0.01; Mass = 200.11;
Mass range = 200.10−200.12
D: Uncertainty = 0.001; Mass = 200.065;
Mass range = 200.064−200.066

  b Each balance may have been calibrated differently.

  c The same balance must be used to minimise random error.

  d Balance D because it weighs to three decimal places.

5 Change in temperature, $\Delta T = 63.73 - 24.15 = 39.58°C$.
Add the uncertainty of each reading = 0.01 + 0.01 = 0.02°C.
The change in temperature is 39.58°C ± 0.02°C.

6 a

Student	Mass (g) ± 0.01 g
Alex	39.35
Ali	39.36
Kim	39.34

  b The students' results were not accurate as the actual value of 38.15 g was not recorded by any student. Their data was precise because all the readings were fairly close to each other. The data was not reliable because each student only recorded one measurement. The three results were all above the actual value indicating a possible systematic error. This may have been due to a poorly calibrated balance or perhaps the students did not zero the balance before using it.

  c average = $\dfrac{39.95 + 39.36 + 39.34}{3} = 39.35\,g$
The largest value of 39.36 g is 0.01 g above the average while the lowest value of 39.34 g is 0.01 g below the average. The absolute uncertainty can be quoted as ± 0.01 g.
However, since the data set is small, the two thirds rule may be applied; i.e. two-thirds of 0.01 g is 0.007 g.
Therefore, average mass of beaker is 39.35 g ± 0.007 g

7 a 3   **c** 1   **e** 2 or 3
  b 3   **d** 2   **f** 3

8 a 0.135 × 202.12 = 27.2862 = 27.3 (to 3 sig fig)
  b 135.62 + 51.1 + 42.367 = 229.087 = 229.1 (to 1 d.p.)

1 a Volume of hydrogen gas produced per minute when Mg is reacted with HCl(aq)

Time (min)	Volume (mL) (± 0.01)
0	0
1	0.5
2	1.0
3	1.6
4	2.3
5	3.0
6	3.5
7	4.0

b

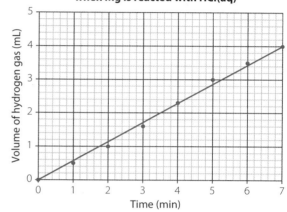

Volume of hydrogen gas produced per minute when Mg is reacted with HCl(aq)

2 a

School day	Temperature (°C)
1	55
2	40
3	5.0
4	39
5	5.0
6	38
7	4.0
8	45
9	55
10	3.0
11	0

  b Average temperature =
$\dfrac{55+40+5+39+5+38+4+45+55+3+0}{11} = 26°C$
The average temperature recorded over the period by the student was 26°C.

  c The limitation of using average values is that there may be some high and some very low readings that skew the results.

  d 55°C and 5.0°C (both of these appear twice)

  e 38°C (the middle value)

3

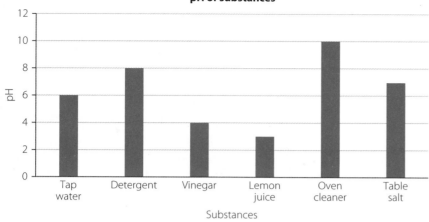

pH of substances

1

V(measuring cylinder) (mL)	V(measured) (mL)	Absolute uncertainty (mL)	% uncertainty
10	10	±0.2	$\frac{0.2}{10} \times 100 = \pm 2$
25	10	±0.25	$\frac{0.25}{10} \times 100 = \pm 2.5$
50	10	±0.5	$\frac{0.5}{10} \times 100 = \pm 5$
100	10	±1	$\frac{1}{10} \times 100 = \pm 10$

2 Conversion of absolute uncertainty to percentage uncertainty:

% uncertainty in mass $= \frac{0.01}{1.45} \times 100 = 0.69\%$

% uncertainty in volume $= \frac{0.25}{250} \times 100 = 0.10\%$

Addition of % uncertainties:

$0.69\% + 0.10\% = 0.79\%$

Conversion of % uncertainty to absolute uncertainty:

Concentration $= \frac{1.45}{0.250} = 5.80 \, g \, L^{-1}$

Absolute uncertainty $= 0.79\%$ of $5.80 = 0.04582... = 0.046$

Concentration of NaCl $= 5.80 \pm 0.046 \, g \, L^{-1}$

3 a

	Reaction time of magnesium metal with different volumes of hydrochloric acid				
Test Tube	Mass of test tube + beaker (±0.001 g)	Mass of test tube + beaker + Mg (±0.001 g)	Mass of Mg (±0.002 g)	V (±0.2 mL)	Time (±0.01 s)
V	75.079	75.105	0.026	2.0	29.49
W	65.669	65.695	0.026	4.0	18.72
X	62.558	62.584	0.026	6.0	12.46
Y	74.745	74.771	0.026	8.0	7.83
Z	74.508	74.534	0.026	10.0	5.01

b **i** What is the effect on the time taken for the same mass of magnesium to disappear when reacted with different volumes of $4 \, mol \, L^{-1}$ hydrochloric acid?

ii If the volume of the $4 \, mol \, L^{-1}$ hydrochloric acid is increased, then the time taken for the magnesium to disappear will decrease.

c Volume of $4 \, mol \, L^{-1}$ hydrochloric acid added

9780170449564

d Time taken for the magnesium strip to disappear

e

Time taken for magnesium to disappear in 4 mol L⁻¹ HCl(aq)

f The student's hypothesis was supported because the time taken for the magnesium to disappear decreased as the volume of hydrochloric acid increased.

g The data collected was not reliable because there were no repeat measurements for each volume of acid.

h The time taken for the magnesium to disappear decreased as the volume of hydrochloric acid was increased in 2 mL increments from 2 mL ± 0.2 mL to 10 mL ± 0.2 mL.

4 a The graphs show scatter plots. They are drawn to see if there is a relationship between variables.

b Graph B is a better representation of the data than graph A. It includes more of the raw data points in the scatter plot and a line of best fit has been drawn to show the trend in the data. A relationship can be seen between temperature and time. Graph A does not have all the experimental data plotted; therefore, a trend cannot be seen. Additionally, graph B has properly labelled axes, and more appropriate intervals on both axes.

5 Osmium will have the least uncertainty in her measured volume of water. Osmium will have to use the 50 mL measuring cylinder once: therefore, the final uncertainty will be 40 mL ± 3 mL.

~continued in right column ▲

Beryllium and fluorine will both have the same uncertainty in their final volume.

Beryllium will have to use the 10 mL measuring cylinder 4 times; therefore, the final uncertainty will be 40 mL ± 4 mL.

Fluorine will have to use the 25 mL measuring cylinder twice; therefore, the final uncertainty will also be 40 mL ± 4 mL.

6 The volume can be anywhere between 14.50 and 15.50 mL when measured using the 50 mL measuring cylinder. When placed in the 25 mL measuring cylinder, the accuracy of the volume will be between 14.45 mL (i.e. 14.50 mL − 0.05 mL = 14.45 mL) and 15.55 mL (i.e. 15.50 mL + 0.05 mL = 15.55 mL).

WS 1.6 PAGE 18

1 a i Drinking 30 mL of Oxydrate is equivalent to drinking 600 mL of pure water.

ii There is no evidence provided for the claim as 600 mL of pure water was not consumed by any participants in order to compare it to the Oxydrate. (Pure water is also not normally consumed.)

b The claim is false. Oxygen has a boiling point of −183°C; therefore, it would not be a liquid at room temperature but instead would be a gas. The claim is that Oxydrate is pure oxygen and it is a liquid at room temperature but the liquid cannot be pure oxygen at room temperature.

c The data is not reliable as the sample size is very small and was not chosen at random, with only five male athletes being given the Oxydrate. The study was conducted over a week but only one set of data is provided and the period for the data not stated. The data collected is not valid for a number of reasons, including: there were no controls placed on what the participants could have for breakfast, including other drinks; the type of activity carried out during the day was not controlled; and feeling thirst is subjective and qualitative.

d The statement is correct. However, the average time for not being thirsty in the test and placebo groups was the same at 3.30 hours; therefore, it is not a valid statement.

e

Issue	Improvement
Sample size of 5	Select more people at random; e.g. 100.
18–22 year old male athletes	Select men and women in various age groups with various levels of fitness.
Meals not controlled	All meals and liquids consumed by participants should be controlled; i.e. be the same.
Level of physical activity not controlled	All participants must have the same level of physical activity during the day.
Drink given before breakfast	Drink should be given after participants undertake the same level of physical activity.
Level of thirst	Feeling thirsty is subjective. It is better to measure the volume of water consumed by participants after a set amount of time after physical activity; e.g. after 30 minutes of the same level of physical activity, wait 30 minutes and measure the volume of water consumed.

f i Oxydrate does not contain electrolytes because the solution did not conduct electricity. If it contained an ionic substance, then the solution would have conducted electricity.

ii Oxydrate may contain a carbon-based covalent molecular compound. Once the solvent was evaporated, the white solid residue had a low melting point and turned black, indicating the presence of carbon. The melting point of 146°C indicates it may be glucose.

WS 1.7 PAGE 20

1 Smith, D., Chowdhury, J., Dreon, S. (2020). *Chemistry in Focus Skills and Assessment Workbook Year 11*. Melbourne: Nelson.

2 a Method

A 100 mL measuring cylinder was used to measure 100 mL of water, ensuring the bottom of the meniscus was on the line. The water was poured into a 250 mL beaker labelled X. A Bunsen burner was set up with a tripod and a gauze mat placed on it. The 250 mL beaker containing the 100 mL water was placed on the gauze mat. A −10 to 110°C thermometer was positioned in the water in the beaker and attached to a retort stand using a clamp. The Bunsen burner was lit and a blue flame was used to heat the water until it started to boil. The temperature was recorded. Four 250 mL beakers were labelled A to D. Using the 100 mL measuring cylinder, 100 mL of water was measured amd placed in each beaker labelled A to D. Four 50 mL beakers were labelled A to D. The 50 mL beaker labelled A was placed on a top-loading balance and 5.0 g of salt was added to it using a spatula. The 50 mL beakers labelled B to D had 10.0 g, 15.0 g and 20.0 g of salt measured and added, respectively. The salt from the 50 mL beaker labelled A was poured into the water in the 250 mL beaker labelled A.

~continued in right column ▲

A glass stirring rod was used to stir the salt solution. The salt from the 50 mL beakers labelled B to D was added to the water in the corresponding 250 mL beakers. Using the −10 to 110°C thermometer, the boiling point of the salt solutions in beakers A to D was determined, as done initially with beaker X.

b Results

Beaker	Observation		Mass of salt (g) ±0.1 g	Boiling temperature (°C) ±1°C
	Initial	Final		
X	Colourless solution	Bubbles and steam formed	0	99
A	Colourless solution and remained colourless when white solid dissolved	Bubbles and steam formed	5.0	103
B	Colourless solution and remained colourless when white solid dissolved	Bubbles and steam formed	10.0	106
C	Colourless solution and remained colourless when white solid dissolved	Bubbles and steam formed	15.0	109
D	Colourless solution and remained colourless when white solid dissolved	Bubbles and steam formed	20.0	115

c

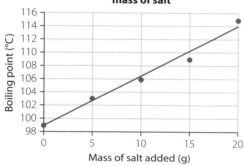

The boiling point of water with varying mass of salt

OR

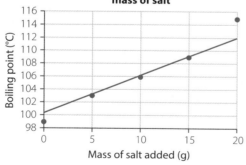

The boiling point of water with varying mass of salt

3 a

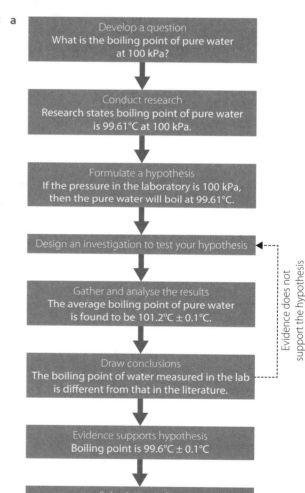

b The student distilled the water first to ensure she was using pure distilled water and she used a thermometer that was calibrated correctly.

9780170449564

MODULE ONE: PROPERTIES AND STRUCTURE OF MATTER

REVIEWING PRIOR KNOWLEDGE PAGE 23

1 A-19, B-17, C-16, D-18, E-20, F-15, G-14, H-13, I-12, J-1, K-2, L-3, M-4, N-5, O-6, P-7, Q-8, R-9, S-10, T-11

2

Mixture	Separation technique
Crude oil	Fractional distillation
Oil and water	Separating funnel
River sand with gold	Sedimentation
Salt water	Distillation
Sand in water	Decantation or filtration
Sawdust and water	Filtration

~continued in right column ▲

3 $\% \text{ salt} = \dfrac{m(\text{salt})}{m(\text{mixture})} \times 100 = \dfrac{12.5}{50} \times 100 = 25.0\%$

$\% \text{ sand} = \dfrac{m(\text{sand})}{m(\text{mixture})} \times 100 = \dfrac{11.6}{50} \times 100 = 23.2\%$

$\% \text{ iron filings} = \dfrac{m(\text{iron filings})}{m(\text{mixture})} \times 100 = \dfrac{25.9}{50} \times 100 = 51.8\%$

4

Name	Formula
Oxygen gas	O_2
Aluminium oxide	Al_2O_3
Carbon dioxide	CO_2
Cobalt	Co
Ammonia	NH_3
Chlorine gas	Cl_2
Silver	Ag

5

Formula	How many different elements?	What are the elements?	Total number of atoms	Made up of ions/molecules
$NaHCO_3$	4	Na, H, C, O	6	ions
$Mg(OH)_2$	3	Mg, O, H	5	ions
Li_3PO_4	3	Li, P, O	8	ions
$KMnO_4$	3	K, Mn, O	6	ions
CCl_4	2	C, Cl	5	molecules

Chapter 2: Properties of matter

WS 2.1 PAGE 25

1 a

Substance	Appearance	Malleable/brittle	Melting point (°C)	Electrical conductivity (S m^{-1})	Classification
W	Shiny, salmon pink	Malleable	1083	5.9×10^7	Metal
X	Dull, grey black	Brittle	3730	1.0×10^5	Non-metal
Y	Dull, yellow	Brittle	98	1.0×10^{-15}	Non-metal
Z	Shiny grey	Brittle	1414	1.0×10^3	Semi-metal

b W can be used for electrical wiring because it has a high electrical conductivity and is ductile.

2

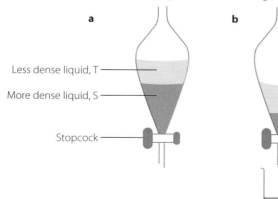
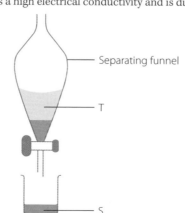

a — Less dense liquid, T; More dense liquid, S; Stopcock

b — Separating funnel; T; S

3 a The mixture is heterogenous. This is because it is composed of solids and liquids that do not dissolve in one another; therefore, it will have non-uniform composition. For example, solid A does not dissolve in C or D and liquids C and D are immiscible.

b

```
                    ┌─────────────────────┐
                    │  Mixture of A, B, C, D │
                    └─────────────────────┘
                               │
              Residue      ◇ Filtration ◇      Filtrate
         ┌─────────────────┘           └─────────────────┐
   ┌───────────┐                                   ┌───────────┐
   │     A     │                                   │  B, C and D │
   └───────────┘                                   └───────────┘
                                                          │
                          Bottom layer  ◇ Separation   ◇  Top layer
                                          using separating
                                          funnel – two
                                          layers form
                         ┌────────────────┘        └────────────────┐
                   ┌───────────┐                              ┌───────────┐
                   │  B and C  │                              │     D     │
                   └───────────┘                              └───────────┘
                          │
                    ◇ Distillation ◇
             Residue                Distillate
        ┌──────────┘                     └──────────┐
   ┌───────────┐                              ┌───────────┐
   │     B     │                              │     C     │
   └───────────┘                              └───────────┘
```

4 a

Groups 1–18

Main group 1, 2 and 13–18

Transition elements – groups 3–12

	1	2	3	4	5	6	7	8	9	10	11	12	13	14	15	16	17	18
Period 1	1 H																	2 He
Period 2	3 Li	4 Be											5 B	6 C	7 N	8 O	9 F	10 Ne
Period 3	11 Na	12 Mg											13 Al	14 Si	15 P	16 S	17 Cl	18 Ar
Period 4	19 K	20 Ca	21 Sc	22 Ti	23 V	24 Cr	25 Mn	26 Fe	27 Co	28 Ni	29 Cu	30 Zn	31 Ga	32 Ge	33 As	34 Se	35 Br	36 Kr
Period 5	37 Rb	38 Sr	39 Y	40 Zr	41 Nb	42 Mo	43 Tc	44 Ru	45 Rh	46 Pd	47 Ag	48 Cd	49 In	50 Sn	51 Sb	52 Te	53 I	54 Xe
Period 6	55 Cs	56 Ba	57-71	72 Hf	73 Ta	74 W	75 Re	76 Os	77 Ir	78 Pt	79 Au	80 Hg	81 Tl	82 Pb	83 Bi	84 Po	85 At	86 Rn
Period 7	87 Fr	88 Ra	89-103	104 Rf	105 Db	106 Sg	107 Bh	108 Hs	109 Mt	110 DS	111 Rg	112 Cn	113 Nh	114 Fi	115 Mc	116 Lv	117 Ts	118 Og

58 La	58 Ce	59 Pr	60 Nd	61 Pm	62 Sm	63 Eu	64 Gd	65 Tb	66 Dy	67 Ho	68 Er	69 Tm	70 Yb	71 Lu
89 Ac	90 Th	91 Pa	92 U	93 Np	94 Pu	95 Am	96 Cm	97 Bk	98 Cf	99 Es	100 Fm	101 Md	102 No	103 Lr

9780170449564

b

	Description	Answer
i	Period 2, group 2	Be or beryllium
ii	Very reactive metal in period 4	K or potassium
iii	Most metallic character in group 1	Fr or francium
iv	The period(s) with only one semi-metal	2, 3
v	The period(s) with only gases	1
vi	The group with alkali metals	1
vii	The group with alkaline earth metals	2
viii	The group with the halogens	17
ix	The group with the noble gases	18

5

Name	Formula
Copper(I) oxide	Cu_2O
Copper(II) sulfate	$CuSO_4$
Sodium carbonate	Na_2CO_3
Magnesium hydroxide	$Mg(OH)_2$
Iron(III) oxide	Fe_2O_3
Sulfur dichloride	SCl_2
Carbon monoxide	CO
Dinitrogen pentoxide	N_2O_5

WS 2.2 PAGE 28

1 a
1. Record the mass of an empty 100 mL beaker.
2. Add the mixture to the beaker and record the mass.
3. Record the mass of a bar magnet.
4. Use the bar magnet to separate the iron filings. Record the mass of the magnet and iron filings.
5. Add sufficient water to the remaining solid sample while stirring.
6. Record the mass of a piece of filter paper.
7. Filter the sample and, using distilled water, rinse the stirring rod from step 4 as well as the residue.
8. Dry the residue, the sand, to constant mass in a drying oven. Record the mass of the filter paper and residue.
9. Record the mass of an evaporating basin.
10. Evaporate the water from the filtrate to dryness to recover the salt. Dry to constant mass. Record the mass of the evaporating basin and filtrate.

b i $m(\text{mixture}) = 99.4\,\text{g} \pm 0.1\,\text{g} - 42.1\,\text{g} \pm 0.1\,\text{g} = 57.3\,\text{g} \pm 0.2\,\text{g}$
$m(\text{iron filings}) = 8.0\,\text{g} \pm 0.1\,\text{g} - 2.1\,\text{g} \pm 0.1\,\text{g} = 5.9\,\text{g} \pm 0.2\,\text{g}$
$m(\text{sand}) = 49.5\,\text{g} \pm 0.1\,\text{g} - 2.2\,\text{g} \pm 0.1\,\text{g} = 47.3\,\text{g} \pm 0.2\,\text{g}$
$m(\text{salt}) = 127.8\,\text{g} \pm 0.1\,\text{g} - 123.7\,\text{g} \pm 0.1\,\text{g} = 4.1\,\text{g} \pm 0.2\,\text{g}$

ii $\%\text{ iron filings} = \dfrac{m(\text{iron filings})}{m(\text{mixture})} \times 100 = \dfrac{5.9}{57.3} \times 100 = 10.30\%$
$= 10.3\%$ (to 3 sig fig)
%uncertainty mass =
$\dfrac{\text{absolute uncertainty}}{\text{mass}} \times 100 = \dfrac{0.2}{5.9} \times 100 = 3\%$

$\%\text{ sand} = \dfrac{m(\text{sand})}{m(\text{mixture})} \times 100 = \dfrac{47.3}{57.3} \times 100 = 82.54\%$
$= 82.5\%$ (to 3 sig fig)
%uncertainty mass =
$\dfrac{\text{absolute uncertainty}}{\text{mass}} \times 100 = \dfrac{0.2}{47.3} \times 100 = 0.4\%$

$\%\text{ salt} = \dfrac{m(\text{salt})}{m(\text{mixture})} \times 100 = \dfrac{4.1}{57.3} \times 100 = 7.155\%$
$= 7.16\%$ (to 3 sig fig)
%uncertainty mass =
$\dfrac{\text{absolute uncertainty}}{\text{mass}} \times 100 = \dfrac{0.2}{4.1} \times 100 = 5\%$

2 $\%\text{ haematite} = \dfrac{100000}{1000000} \times 100 = 10\%$

$\%\text{ magnetite} = \dfrac{990}{1000000} \times 100 = 0.099\%$

3 a $m(\text{pseudoephedrine hydrochloride}) = \dfrac{1.1}{100} \times 5.2\,\text{g} = 0.057\,\text{g}$
(to 2 sig fig)

b maximum number of tablets $= \dfrac{240\,\text{mg}}{57\,\text{mg}} = 4.2\ldots = 4$

The maximum number of tablets that can be consumed in 24 hours is four.

4 $\text{ppm} = \dfrac{\text{mg}}{\text{kg}}$

$570\,000 = \dfrac{\text{mass of copper in mg}}{2\,\text{kg}}$

Mass of copper in mg $= 570\,000 \times 2 = 1\,140\,000\,\text{mg}$
Mass of copper in kg $= 1.14\,\text{kg}$

$\%\text{ copper in dumbbell} = \dfrac{1.14}{2} \times 100 = 57\%$

Since the copper content of the dumbbell is 57%, which is less than the 65% required, it is not suitable to reduce bacteria.

Chapter 3: Atomic structure and atomic mass

WS 3.1 PAGE 30

1 An atom is made up of a nucleus that is **small** and **dense** because it holds protons and **neutrons**. The protons have a **positive** charge while the electrons have a **negative** charge. The nucleus has **a positive** charge because **it has positively charged protons and neutrons which have no charge**. The nucleus is surrounded by a cloud of **electrons**. The bulk of the volume of the atom is from the **electron cloud**. The nucleus has a diameter **which is less than one ten thousandth** times the diameter of the whole atom. An element is made up of **the same type** of atoms. The atomic number, with symbol **Z**, refers to the number of **protons** while the mass number, with symbol **A**, refers to the number of protons **and neutrons**. The term 'nucleon' is used to describe either a proton or **a neutron**. The **mass** number is sometimes called the nucleon number. Isotopes of elements are atoms that have different numbers of **neutrons** in their nuclei. Isotopes of the same element will have the same **atomic** number but different **mass** numbers.

2

Isotope	Symbol	Atomic number	Mass number	Number of protons	Number of neutrons	Number of electrons
Iodine-125	I	53	125	53	72	53
Iodine-131	I$^-$	53	131	53	78	54

3 unstable; spontaneously; radiation; radioisotopes; unstable; stability; high; unstable; less; equal; 1; not; unstable; 6; neutron; 1; 6; 7; 6; 7; not; unstable; increases; stable; greater; greater; less; not; unstable; unstable; 125; 1:5; stable; 132; 82; 1:6; unstable; greater; unstable; alpha; protons

4

Isotope	Neutron to proton ratio or $Z > 83$, too big	Stable or unstable
$^{3}_{1}$H	$2:1 \neq 1$	Unstable
$^{20}_{10}$Ne	$10:10 = 1$	Stable
$^{232}_{90}$Th	$Z > 83$, too big	Unstable
$^{37}_{17}$Cl	$20:17 \neq 1$	Unstable
$^{231}_{91}$Pa	$Z > 83$, too big	Unstable
$^{197}_{79}$Au	$118:79 = 1.49 = 1.5$	Stable
$^{18}_{8}$O	$10:8 \neq 1$	Unstable
$^{238}_{92}$U	$Z > 83$, too big	Unstable

5 a Elements with an atomic number 1–20: hydrogen, helium, lithium, beryllium, boron, carbon, nitrogen, oxygen, fluorine, neon, sodium, magnesium, aluminium, silicon, phosphorus, sulfur, chlorine, argon, potassium, calcium.

~continued in right column ▲

b Elements with atomic number greater than 83: polonium, astatine, radon, francium, radium, actinoids (actinium, thorium, protactinium, uranium, neptunium, plutonium, americium, curium, berkelium, californium, einsteinium, fermium, mendelevium, nobelium, lawrencium), rutherfordium, dubnium, seaborgium, bohrium, hassium, meitnerium, darmstadtium, roentgenium, copernicium, nihonium, flerovium, moscovium, livermorium, tennessine, oganesson.

WS 3.2 PAGE 33

1 a

	Term	Label number
i	d	10
ii	s	6
iii	p_x	7
iv	p_y	8
v	p_z	9
vi	Orbital	5
vii	Main energy level	11
viii	Energy level 1	1
ix	Energy level 2	2
x	Energy level 3	3
xi	Nucleus	4

b

Main energy level	Number of sublevels	Identity of sublevels	Maximum number of electrons
1	1	1s	2
2	2	2s and 2p	8
3	3	3s, 3p and 3d	18
4	4	4s, 4p, 4d and 4f	32

c The maximum number of electrons an orbital can hold is two.

d The diameter of the spheres increases from 1s to 3s with an increase in energy level.

2 a i $1s^2 2s^2 2p^3$
 ii $1s^2$
 iii $1s^2 2s^2 2p^6 3s^2 3p^6 4s^2$
 iv $1s^2 2s^2 2p^6 3s^2 3p^6$

b The species may be: N^{3-}, O^{2-}, F^-, Ne, Na^+, Mg^{2+} or Al^{3+}

3 A = Absorption spectrum

 B = Emission spectrum

4 a The high voltage provides energy to excite the electrons from their ground state to their excited state. However,

~continued in right column ▲

when these electrons fall back to their ground state after a short time, the excess energy is emitted as light at various wavelengths.

b The spectroscope or prism breaks up the emitted light into its various wavelength components. The emissions have actually occurred at just a few discrete (separate or distinct) wavelengths; however, we see the combination of these distinct emissions as a purple-pink colour.

c Each line is specific to a particular wavelength and corresponds the particular quantity of energy required to excite a specific electron in the atom from its ground state to an excited state. An equal quantity of energy is released when that electron falls back from the excited to the ground state.

9780170449564

d The shorter the wavelength, the greater the amount of energy released, hence the bigger the jump.

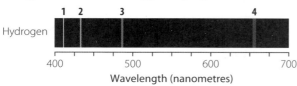

Hydrogen

Wavelength (nanometres)

5 a Lithium and sodium

b The flame colour for the mixture would most likely be an orange because lithium produces a red flame, while sodium produces a yellow flame.

WS 3.3 PAGE 37

1 **i** Alpha (α) ray; **ii** Beta (β) ray; **iii** Gamma (γ) ray

2 The identity of the radioisotope is 'd'. The radioisotope has to be a beta or gamma emitter because the paper would have stopped alpha particles. The time difference means the rate had dropped to one-quarter of its original value; i.e. 2 half-lives had passed in 16 days. Therefore, the half-life must be 8 days. Radioisotope 'd' is a beta emitter with a half-life of 8 days.

3 a $^{214}_{84}\text{Po} \rightarrow {}^{4}_{2}\text{He} + {}^{210}_{82}\text{Pb}$

b $^{14}_{6}\text{C} \rightarrow {}^{0}_{-1}e + {}^{14}_{7}\text{N}$

c $^{238}_{92}\text{U} + {}^{1}_{0}n \rightarrow {}^{239}_{92}\text{U} \rightarrow {}^{0}_{-1}e + {}^{239}_{93}\text{Np}$

d $^{239}_{94}\text{Pu} + 2\ {}^{1}_{0}n \rightarrow {}^{241}_{94}\text{Pu} \rightarrow {}^{241}_{95}\text{Am} + {}^{0}_{-1}e$

4 a It is desirable to use radioisotopes with a shorter half-life for diagnosis to minimise patient exposure. Therefore, iodine-123 is better than iodine-131, as it has a half-life of 13 hours and emits only gamma radiation. The shorter half-life means minimal exposure for the patient and gamma rays can pass straight through the body with minimal damage to tissue, so ideal for diagnosis. Iodine-131 emits beta radiation, which can kill the cancer cells when treating a patient. It also has a longer half-life of 8 days, which increases the patient's exposure to the radiation to help kill the cancer cells.

b $^{131}_{53}\text{I} \rightarrow {}^{131}_{54}\text{Xe} + {}^{0}_{-1}e + \gamma$

5 a $^{226}_{88}\text{Ra} \rightarrow {}^{4}_{2}\text{He} + {}^{222}_{86}\text{Rn}$

b The alpha radiation may have affected the fine corneal tissue in the eye and may have caused mutations in proteins and other molecules in the body.
The beta radiation may have killed healthy cells.

c Since radium-226 has a half-life of 1600 years, the watches with the radium dials are still radioactive as they have not even undergone one half-life. The watches should have been disposed of in special storage that would have allowed for the expansion of volume during the decay process due to the formation of radon, which is a gas at room temperature.

Chapter 4: Periodicity

WS 4.1 PAGE 39

1

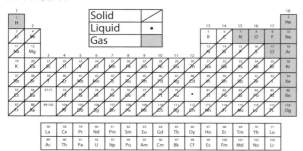

---

2 a

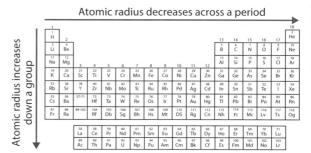

b Across a period, the atomic radius decreases because as the atomic number increases the number of protons in the nucleus increases, and the electrons are added to the same main energy level; hence, the force of attraction between the nucleus and each electron in the outermost shell is increasing, drawing the electron closer to the nucleus, thus making the radius smaller.

Going down a group, atomic radius increases because as the atomic number increases the number of protons in the nucleus increases, and the electrons are now being added to higher main energy levels. The outermost electrons are further away from the nucleus and also experience the shielding effect from completely filled shells closer to the nucleus, which means there is a decrease in the electrostatic force of attraction between the nucleus and the outermost electrons.

c Both the sodium ion and the oxide ion have the same electron configuration of $1s^2 2s^2 2p^6$. However, since sodium ion has 11 protons while oxide ion has 8 protons, the electrostatic force of attraction from the nucleus of the sodium ion will be greater, making its radius smaller than the oxide ion.

3 a Ionisation energy is the energy required to remove an electron from a gaseous atom. The unit for ionisation energy is J per atom or, more commonly, kilojoules per mole of atoms, kJ mol^{-1}.

b

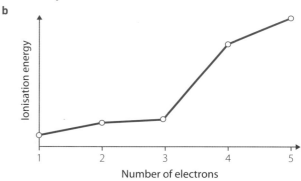

c 1st ionisation: $\quad \text{B}(g) \rightarrow \text{B}^+(g) + \text{e}^-$
2nd ionisation: $\quad \text{B}^+(g) \rightarrow \text{B}^{2+}(g) + \text{e}^-$
3rd ionisation: $\quad \text{B}^{2+}(g) \rightarrow \text{B}^{3+}(g) + \text{e}^-$
4th ionisation: $\quad \text{B}^{3+}(g) \rightarrow \text{B}^{4+}(g) + \text{e}^-$
5th ionisation: $\quad \text{B}^{4+}(g) \rightarrow \text{B}^{5+}(g) + \text{e}^-$

d **i** $1s^2 2s^2 2p^1$

ii 1st ionisation: one electron from energy level 2 and a *p* orbital

2nd ionisation: one electron from energy level 2 and *s* orbital

3rd ionisation: another electron from energy level 2 and *s* orbital

4th ionisation: one electron from energy level 1 and *s* orbital

5th ionisation: another electron from energy level 1 and *s* orbital

4 a Electronegativity of an element is a measure of the ability of an atom of that element to attract bonding electrons towards itself in compounds.

  b Electronegativity increases across a period, and decreases down a group, excluding the noble gases.

  c The Pauling scale is used to assign the values for electronegativity.

5 a Group 1

  b Group 2

  c Magnesium (also Al, Zn, Fe)

  d True

  e Sodium hydroxide, NaOH

  f Hydrogen gas, H_2

  g False

  h False

6 a Chlorine and fluorine, because they react with water.

  b The collected gas may be oxygen because it re-ignited a glowing splint. However, it could also be fluorine that has reacted with water to produce oxygen gas. Before bubbling gas X through water, Jade must first test the water with litmus to ensure it is neutral at the start of the experiment. If the litmus does not change colour after gas X is bubbled through, then gas X is oxygen. However, if the litmus changes to red after gas X is bubbled through, then gas X must be fluorine because it produces HF(aq), which is acidic. Jade could also test gas X at the start of the experiment with moistened litmus. If it stays the same, the gas is oxygen; if it turns red, it is fluorine.

Chapter 5: Bonding

WS 5.1 PAGE 43

1 a $\Delta EN(\text{CsCl}) = 3.16 - 0.79 = 2.37$

  $\Delta EN \geq 1.8\text{--}2.0$ Therefore, bonding in CsCl is ionic.

  b $\Delta EN(\text{CaO}) = 3.44 - 1.00 = 2.44$

  $\Delta EN \geq 1.8\text{--}2.0$ Therefore, bonding in CaO is ionic.

  c $\Delta EN(\text{CO}) = 3.44 - 2.55 = 0.89$

  $0 \leq \Delta EN \leq 1.8\text{--}2.0$ Therefore, bonding in CO is polar covalent.

  d $\Delta EN(\text{Cl}_2) = 3.16 - 3.1 = 0$

  $\Delta EN = 0$ Therefore, bonding in Cl_2 is pure covalent.

  e $\Delta EN(\text{HCl}) = 3.16 - 2.20 = 0.96$

  $0 \leq \Delta EN \leq 1.8\text{--}2.0$ Therefore, bonding in HCl is polar covalent.

2 a **i** $2\text{Na(s)} + \text{Cl}_2\text{(g)} \rightarrow 2\,\text{NaCl(s)}$

  ii
  $$2\,\overset{\bullet}{\text{Na}}\,\text{(s)} + \;:\!\overset{\cdot\cdot}{\underset{\cdot\cdot}{\text{Cl}}}\!:\!\overset{\cdot\cdot}{\underset{\cdot\cdot}{\text{Cl}}}\!:\;\text{(g)} \rightarrow 2\Big[\text{Na}\Big]^{+}\Big[:\!\overset{\cdot\cdot}{\underset{\cdot\cdot}{\text{Cl}}}\!:\Big]^{-}$$

  b **i** $2\text{Mg(s)} + \text{O}_2\text{(g)} \rightarrow 2\text{MgO(s)}$

  ii
  $$2\!\cdot\!\text{Mg}\!\cdot\!\text{(s)} + \;:\!\overset{\cdot\cdot}{\text{O}}\!:\!\overset{\cdot\cdot}{\underset{\cdot\cdot}{\text{O}}}\!:\;\text{(g)} \rightarrow \Big[\text{Mg}\Big]^{2+}\Big[:\!\overset{\cdot\cdot}{\underset{\cdot\cdot}{\text{O}}}\!:\Big]^{2-}$$

3 a

Species	Valency
Ag	1
C	4
F	1
He	0
N	3
O	2
Zn	2

b

Metal element	Ions formed	Valency	Name with valency
Copper	Cu^+	1	Copper(I)
	Cu^{2+}	2	Copper(II)
Gold	Au^+	1	Gold(I)
	Au^{3+}	3	Gold(III)
Iron	Fe^{2+}	2	Iron(II)
	Fe^{3+}	3	Iron(III)
Lead	Pb^{2+}	2	Lead(II)
	Pb^{4+}	4	Lead(IV)
Mercury	Hg_2^{2+}	1	Mercury(I)
	Hg^{2+}	2	Mercury(II)
Tin	Sn^{2+}	2	Tin(II)
	Sn^{4+}	4	Tin(IV)

c

Name	Formula
Copper(II) sulfate	$CuSO_4$
Gold(III) chloride	$AuCl_3$
Iron(II) hydroxide	$Fe(OH)_2$
Mercury(II) oxide	HgO
Lead(IV) iodide	PbI_4
Tin(II) sulfide	SnS
Mercury(I) bromide	Hg_2Br_2
Copper(I) oxide	Cu_2O

4 a

Compound	Valency of underlined element
H<u>F</u>	1
<u>N</u>H$_3$	3
H$_2$<u>S</u>	2
<u>C</u>O$_2$	4
<u>P</u>Cl$_5$	5

b

Name	Formula
Dinitrogen pentoxide	N_2O_5
Sulfur hexafluoride	SF_6
Carbon tetrachloride	CCl_4
Sulfur dioxide	SO_2
Sulfur tetrafluoride	SF_4
Nitrogen trichloride	NCl_3
Ammonia	NH_3
Silicon dioxide	SiO_2

9780170449564

1

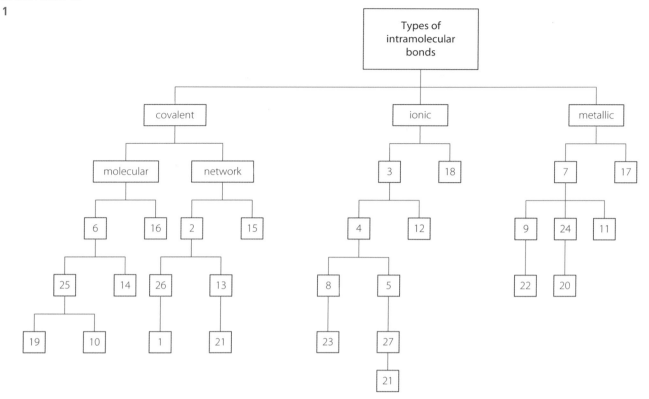

2

Description	Term
Sulfur appears …	dull
Copper appears …	lustrous
Carry the charge in graphite	electrons
Carry the charge in molten NaCl	ions
Hydrogen molecule is made up of …	atoms
Methane molecule has …	temporary dipoles
Hydrogen sulfide has …	permanent dipoles
Is made up of ions	KCl
Can conduct electricity in the solid form	Ag
Magnesium oxide is …	brittle

WS 5.3 PAGE 48

1 a   Carbon dioxide: $\overset{..}{O}::C::\overset{..}{O}$ Shape – linear

b   Ammonia: $H:\overset{..}{N}:H$ Shape – pyramidal
  H

c   Carbon tetrachloride:
  Shape – tetrahedral

d   Phosphorus pentachloride:
  Shape – trigonal bipyramidal

e   Boron trifluoride:
  Shape – trigonal planar

f   Hydrogen sulfide:
  Shape – bent

2   Carbon dioxide is linear while hydrogen sulfide is a bent molecule. According to the valence shell electron-pair repulsion theory, the repulsion between a lone pair and a bonded pair is greater than between bonded pairs. The central atom in carbon dioxide, carbon, has four bonded pairs, while the central atom in hydrogen sulfide has two lone pairs of electrons. Repulsion between bonded pairs on opposite sides of the carbon atom is much less than the repulsion between the lone pairs on sulfur and the two bonded pairs on either side of the sulfur atom. Thus, the repulsive force of the lone pairs on the bonded pairs pushes the bonded pairs together, making the hydrogen sulfide molecule bent, while carbon dioxide is linear.

3 a   element; element; covalently; bonded; different; state; gaseous; linear; bent; diamond; tetrahedrally; graphite; planar; valence; conduct; flat; parallel; buckminsterfullerene; nanotubes

b   i   Diamond
  ii   Nanotube
  iii   Graphite
  iv   Buckminsterfullerene

c   i   Hardness: Very hard
    Explanation: The three-dimensional lattice structure with strong covalent bonds make diamond very hard.
    Electrical conductivity: Does not conduct electricity.
    Explanation: All four of carbon's valence electrons are tied up in forming the four covalent bonds. So, there are no mobile electrons to conduct electricity.

ii Hardness: Hard

Explanation: Depends on how the sheets of graphite are rolled into cylinders to make nanotubes.

Electrical conductivity: Conducts electricity

Explanation: Each carbon is bonded to three other carbons, leaving the fourth valence electron mobile.

iii Hardness: Soft

Explanation: Planar structure of six-membered rings in layers with weak intermolecular forces between the layers. Therefore, the layers can move, making graphite soft.

Electrical conductivity: Conducts electricity

Explanation: Each carbon is bonded to three other carbons, leaving the fourth valence electron mobile. Therefore, it conducts electricity.

d

Allotrope	Use	Related property
Diamond	Drill bit	Very hard
Graphite	Industrial lubricant	Weak intermolecular forces between the layers means they can easily slide over one another
Buckminsterfullerene	Medicine	Can deliver drugs, trapped in a nano cage, to sites in the body
Nanotube	Wiring for electrical circuits	Good conductors of electricity

WS 5.4 PAGE 51

1 between; physical; stronger; higher; type; non-polar; non-polar; polar; symmetrical; weakest; strength; size; shape; larger; less; greater; stronger; polar; asymmetrical; strongest

2 a H_2O: Hydrogen bonds

Hydrogen bond

$$H^{\delta+} \qquad\qquad H^{\delta+}$$
$$O_{\delta-} - - - H_{\delta+} - O_{\delta-}$$
$$H_{\delta+}$$

b HCl: Dipole–dipole forces

Dipole–dipole forces

$$\overset{\delta+}{H} - \overset{\delta-}{Cl} - - - \overset{\delta+}{H} - \overset{\delta-}{Cl}$$

c ICl: Dipole–dipole forces

Dipole–dipole forces

$$\overset{\delta+}{I} - \overset{\delta-}{Cl} - - - \overset{\delta+}{I} - \overset{\delta-}{Cl}$$

d Cl_2: Dispersion forces

$$Cl - Cl$$
$$Cl - Cl \qquad Cl - Cl$$
$$Cl - Cl \qquad Cl - Cl$$

Dispersion forces

3 The stronger the strength of the intermolecular force, the higher the boiling point. Pentane has a higher boiling point (36°C) than 2,2-dimethylpropane (10°C) because more energy is required to separate the neighbouring molecules of pentane. Pentane molecules can pack close together compared to the branched 2,2-dimethylpropane, which is not able to pack as closely together, so less energy is required to separate these molecules.

4

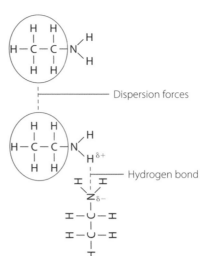

Dispersion forces

Hydrogen bond

MODULE ONE: CHECKING UNDERSTANDING PAGE 53

1 A-6, B-16, C-17, D-8, E-4, F-13, G-14, H-7, I-18, J-11, K-5, L-3, M-15, N-1, O-12, P-2, Q-10, R-20, S-19, T-9

2 a Fractionating column

b Separating funnel

c Filter funnel

d Condenser

e Evaporating basin

f Round-bottom flask

g Tripod and gauze mat

h Conical flask

3 1 Filter the mixture.

2 The residue will be Y.

3 Fractionally distil the filtrate by using a heating mantle. A naked flame must not be used as the X and Z components are volatile.

4 The distillate that is captured first will be X as it has a lower boiling point than Z, and the second distillate will be Z.

4 a Carbon, hydrogen, nitrogen, oxygen

b Mass % C = $\dfrac{168.14}{294.304} \times 100 = 57.14\%$

Mass % H = $\dfrac{18.144}{294.304} \times 100 = 6.165\%$

Mass % N = $\dfrac{28.02}{294.304} \times 100 = 9.521\%$

Mass % O = $\dfrac{80.00}{294.304} \times 100 = 27.18\%$

Add the % together to ensure they add up to 100%: 57.14% + 6.165% + 9.521% + 27.18% = 100.006% = 100%

5

Species	Electron configuration (*spdf* notation)
Carbon	$1s^2 2s^2 2p^2$
Chromium	$1s^2 2s^2 2p^6 3s^2 3p^6 3d^5 4s^1$
Copper	$1s^2 2s^2 2p^6 3s^2 3p^6 3d^{10} 4s^1$
Fluoride	$1s^2 2s^2 2p^6$
Lithium	$1s^2 2s^1$
Magnesium	$1s^2 2s^2 2p^6 3s^2$
Oxide ion	$1s^2 2s^2 2p^6$

9780170449564

6 a i d

    ii p

    iii s

    iv f

b The maximum number of electrons one orbital can hold is 2.

7 a Ionic; e.g. NaCl

b Covalent (molecular); e.g. HCl

c Covalent (network); e.g. diamond

d Metallic; e.g. Mg or any metal

8 a $^{238}_{92}U \rightarrow {}^{4}_{2}He + {}^{234}_{90}Th$

    $^{234}_{90}Th \rightarrow {}^{0}_{-1}e + {}^{234}_{91}Pa$

b

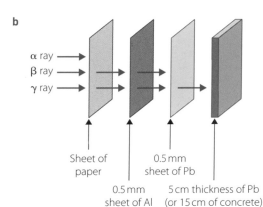

α ray
β ray
γ ray

Sheet of paper

0.5 mm sheet of Pb

0.5 mm sheet of Al

5 cm thickness of Pb (or 15 cm of concrete)

9 a The coloured lines represent wavelength of light emitted. They correspond to the energy required to excite a particular electron in the atom from its ground state to an excited state, which is equal to the energy released when that electron falls from the excited state to the ground state. The shorter the wavelength of the radiation, the greater the amount of energy released. Each of the energy sublevels, such as the $2p$, $3p$ or $3d$, has a different energy in the atoms of different elements. Hence, each element has its unique emission spectrum.

b The Bohr theory was that electrons moved around the nucleus in fixed orbits. When they absorb energy, they move into orbits of larger radius and when they emit energy they fall back to smaller orbit. This basis was used to explain the spectrum of hydrogen, however, it could not be used to quantitatively interpret more complex spectra. Schrödinger, on the other hand, treated the electron in the hydrogen atom as a wave instead of as a particle. He developed mathematical equations to determine the probability of finding an electron in a volume of space around the nucleus. The Schrödinger equation was successfully used to interpret emission spectra of atoms with many electrons because each line in the spectrum corresponds to a specific decrease in energy level.

10 a Atomic radius decreases from the sodium atom to the chlorine atom. This is because the number of protons in the nucleus is increasing across the period while electrons are being added to the same shell. The stronger electrostatic pull from the nucleus results in a smaller atomic radius.

b Both the sodium ion and the oxide ion have the same number of electrons; however, the sodium ion has 11 protons in its nucleus while the oxide ion has 8 protons. Therefore, the nuclear pull on the electrons will be greater for the sodium ion, making its radius smaller than the radius of the oxide ion.

c i The formula of the compound formed between X and Y is X_3Y_2. There is a big jump between the second and third ionisation for X, while there is a big jump between fifth and sixth for Y. These jumps show that X has two valence electrons while Y has five valence electrons. X is likely to be a metal and Y a non-metal. Therefore, X would form a cation with charge $+2$ while Y would form an anion with charge -3. The ionic compound formed between X and Y would therefore have the formula X_3Y_2.

ii X has 12 electrons and Y has 15 electrons. Therefore, X has 12 protons with an atomic number of 12 and Y has 15 protons with an atomic number of 15. Therefore, X is magnesium and Y is phosphorus.

11 a Melting point is determined by the strength of intermolecular forces. Since the melting point of X is lower than Y, X may be a covalent molecular substance with weak intermolecular forces while Y may be an ionic substance with strong electrostatic forces of attraction between ions in the lattice. Therefore, more energy is required to melt Y than is required to melt X.

b X and Y could be dissolved in water and the conductivity of the water tested. The solution of X should not conduct electricity while the solution of Y should conduct electricity.

12 Allotropes are different structural forms of the same element in the same physical state that have distinctly different physical properties. Isotopes on the other hand, are atoms of the one element that have different numbers of neutrons in their nuclei. Isotopes of the one element have the same atomic number but different mass numbers. Examples of allotropes include the element carbon in different structural forms such as diamond, graphite and fullerenes. Examples of isotopes include the element hydrogen as hydrogen 1, H-1, deuterium, H-2, and tritium, H-3.

MODULE TWO: INTRODUCTION TO QUANTITATIVE CHEMISTRY

REVIEWING PRIOR KNOWLEDGE PAGE 61

1 a T, T, F, T, F, F, T, F

b Salt, oxygen gas and anhydrous iron(II) chloride are all pure substances.

Sedimentation can be used to separate wooden blocks from rocks.

The electron configuration of nickel is $1s^2 2s^2 2p^6 3s^2 3p^6 4s^2 3d^8$.

Copper salts burn with a blue-green flame.

2

Chemical name	Chemical formula	Chemical name	Chemical formula
Sodium chloride	NaCl	Barium sulfate	$BaSO_4$
Magnesium oxide	MgO	Magnesium chloride	$MgCl_2$
Sodium carbonate	Na_2CO_3	Lithium phosphide	Li_3P

3 a

$:\!\overset{\cdot}{Al} + 3 \cdot \overset{\cdot\cdot}{\underset{\cdot\cdot}{Cl}}: \rightarrow \left[:\overset{\cdot\cdot}{\underset{\cdot\cdot}{Cl}}:\right]^- Al^{3+} \left[:\overset{\cdot\cdot}{\underset{\cdot\cdot}{Cl}}:\right]^-$

$\left[:\overset{\cdot\cdot}{\underset{\cdot\cdot}{Cl}}:\right]^-$

b $2\overset{\cdot}{H} + \cdot\overset{\cdot\cdot}{\underset{\cdot\cdot}{S}}: \rightarrow H:\overset{\cdot\cdot}{\underset{\cdot\cdot}{S}}:H$

4 The properties of the ionic compound calcium bromide are different from those of the metal calcium and the non-metal bromine. Calcium bromide is white in colour. A mixture of calcium and bromine would display some of the properties of the individual elements – a silver solid within a brown liquid or some parts boiling at 59°C and others at 1484°C. As calcium bromide has a distinct melting and boiling point, not a range, it is a pure substance.

5 a 1 Weigh and record the mass of an empty crucible on an electronic scale and record the mass.

    2 Add 12.5 g of hydrated copper(II) sulfate to the crucible. Weigh and record the gross mass of the crucible and hydrated copper(II) sulfate.

    3 Strongly heat the crucible over a Bunsen burner for 5 minutes. Allow to cool. Weigh and record the gross mass of the crucible and copper(II) sulfate.

    4 Continue to heat the crucible and record the gross mass until constant mass is achieved.

b

What are the risks in doing this investigation?	How can you manage these risks to stay safe?
Spitting copper(II) sulfate may cause burns to eyes and face, and contaminate lab bench.	Wear safety glasses. Secure lid of crucible while heating.
Hot objects may cause burns.	Turn off Bunsen burner between heatings. Allow the crucible to cool completely before weighing. Allow all equipment to cool completely before packing away. Hair to be tied back, loose clothing to be contained within a lab coat.
Copper(II) sulfate is toxic.	Do not allow copper(II) sulfate to contact skin. If contact occurs, wash with plenty of water. Dispose of copper in waste jar. Do not put in the rubbish bin or down the sink.

c Heating until constant mass is achieved ensures that all the water has been evaporated. Had the student halted after three heatings, the experiment could be invalid as water may remain within the sample.

d Theoretical:

Molar mass($CuSO_4.5H_2O$)

$= 63.55 + 32.07 + (16.00 \times 4) + (1.008 \times 10) + (16.00 \times 5)$

$= 249.7\ g\,mol^{-1}$

Mass of H_2O in 1 mol $CuSO_4.5H_2O = (1.008 \times 10) + (16.00 \times 5)$

$= 90.8\ g$

% H_2O in $CuSO_4.5H_2O = \dfrac{90.8}{249.7} \times 100$

$= 36.4\%$

Experimental:

Mass of water = initial mass − final heating mass

$= 80.0 - 76.1$

$= 3.9\ g$

Mass of hydrated copper(II) sulfate

$=$ initial mass − mass of crucible

$= 80.0 - 67.5$

$= 12.5\ g$

Percentage of water

$= \dfrac{\text{(mass of water)}}{\text{(mass of hydrated copper(II) sulfate)}} \times 100$

$= \dfrac{3.9}{12.5} \times 100$

$= 31.2\%$

e

  ⬦ Although the sample was heated to constant mass, water may remain within the sample.

  ⬦ Soot may have accumulated on the bottom of the crucible during heating.

  ⬦ Experimental error of the electronic scale ($\pm\ 0.1g$).

f This experiment is unreliable as it was only performed once on a single sample of hydrated copper (II) sulfate. Heating to constant mass is not repetition of an experiment. Reliability could be improved by heating multiple samples of hydrated copper sulfate, averaging results and excluding outliers.

6 a Ionisation energy generally increases across the period. As you move across the period, the atomic radius decreases, meaning the outer electrons are closer to the nucleus. Therefore, the energy required to remove these electrons increases.

b Exception 1: Al is lower than Mg but, according to the trend, it should be higher.

The electron configuration of Mg is $1s^2 2s^2 2p^6 3s^2$ whereas Al is $1s^2 2s^2 2p^6 3s^2 3p^1$. Therefore, it requires less energy to remove the lone electron from aluminium's $3p^1$ orbital than from magnesium's paired $3s^2$ orbital.

Exception 2: S is lower than P but, according to the trend, it should be higher.

The electron configuration of phosphorus is $1s^2 2s^2 2p^6 3s^2 3p^3$ whereas sulfur is $1s^2 2s^2 2p^6 3s^2 3p^4$. The $3p$ electrons in phosphorus are all unpaired, whereas sulfur has one paired set. This pair slightly repel each other so their force to the nucleus is slightly reduced, making it easier to remove one of these electrons.

7 Isotopes are atoms of the same element that have different numbers of neutrons in their nuclei even though they have the same number of protons.

8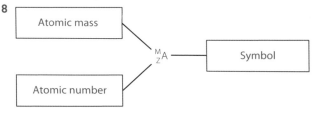

9

Isotope name	Isotopic notation	Atomic number	Number of protons	Number of neutrons
Helium-4	$^{4}_{2}\text{He}$	2	2	2
Boron-11	$^{11}_{5}\text{B}$	5	5	6
Potassium-37	$^{37}_{19}\text{K}$	19	19	18
Americium-241	$^{241}_{95}\text{Am}$	95	95	146
Uranium-235	$^{235}_{92}\text{U}$	92	92	143

Chapter 6: Chemical reactions and stoichiometry

WS 6.1 PAGE 65

1 The law of conservation of mass states that matter can be neither created nor destroyed, but merely changed from one form to another.

2 a calcium chloride + sodium sulfate → calcium sulfate + sodium chloride

b If the mass of the reactants equals 300.23 g, then the mass of the products will equal 300.23 g.

c 1 Put on safety glasses. Have paper towel on hand to clean up any spills straight away.

2 Measure 20 mL of sodium sulfate solution. Pour the sodium sulfate solution into the conical flask.

3 In a clean measuring cylinder, measure 20 mL of calcium chloride solution and pour into the test tube.

4 Carefully place the test tube inside the conical flask. Seal the conical flask with a stopper.

5 Weigh the complete apparatus. Record the mass.

6 Gently invert the conical flask, allowing the two solutions to mix.

~continued in right column ▲

7 Weigh the complete apparatus again. Record the final mass.

8 Repeat this experiment several times, excluding outliers, and average results to obtain a reliable result.

d This experiment determined that the mass of reactants (300.23 g) was equal to the mass of the products (300.23 g). This confirms the hypothesis and the law of conservation of mass.

3 a Trial 2 mass of the beaker + vinegar is approx. 10 g higher than the other trials, resulting in a higher final mass. A mistake or random error may have occurred where the student didn't read the meniscus of the measuring cylinder properly when measuring the volume of vinegar, or the student may have used a different-sized beaker or did not remove their hand from the beaker when recording its mass.

Note that the mass of baking soda in Trial 3 is not an outlier. An electronic balance has an error of ±0.1 g, which means this value is within an acceptable range as the method asks for 5.0 g.

b

	Trial 1	Trial 2	Trial 3	Trial 4	Average (g)
Mass of beaker + vinegar (g)	231	242	235	234	233
Mass of baking soda (g)	4.9	4.9	5.1	4.9	4.9
Mass of beaker + vinegar + baking soda (g)	225	238	226	228	226

c The results do not demonstrate the law of conservation of mass due to an invalid method, rather than a systematic or random error. One of the products of this reaction is the gas carbon dioxide. Performing this experiment in an open beaker allows the carbon dioxide to escape into the atmosphere, resulting in a lower final mass and, therefore, invalidating the experiment.

4 a i Fe(s) + CuCl$_2$(s) → Cu(s) + FeCl$_2$(s)

ii Zn(s) + 2HCl(aq) → ZnCl$_2$(aq) + H$_2$(g)

iii F$_2$(g) + Ca(s) → CaF$_2$(s)

iv 2NaOH(s) + H$_2$SO$_4$(aq) → Na$_2$SO$_4$(aq) + 2H$_2$O(l)

b i 55.85 + 134.45 → 63.55 + 126.75
Total reactants = 190.3; Total products = 190.3

ii 65.38 + 36.458 → 136.28 + 2.016
Total reactants = 101.838; Total products = 138.296

iii 38 + 40.08 → 78.08 Total reactants = 78.08;
Total products = 78.08

iv 39.998 + 98.086 → 142.05 + 18.016
Total reactants = 138.084; Total products = 160.066

5 The law of conservation of mass states that matter can be neither created nor destroyed, but merely changed from one

~continued in right column ▲

form to another. As seen in equations ii and iv, the mass of the reactants and mass of the products do not equate. Equations need to be balanced for the number of each type of atom, and hence the mass, to be the same on both sides.

ii 65.38 + (2 × 36.458) → 136.28 + 2.016
Total reactants = 138.296; Total products = 138.296

iv (2 × 39.998) + 98.086 → 142.05 + (2 × 18.016)
Total reactants = 178.082; Total products = 178.082

Chapter 7: Mole concept

WS 7.1 PAGE 68

1 a Relative atomic mass is the average mass of atoms present in the naturally occurring element relative to the mass of an atom of the carbon-12 isotope taken as exactly 12.

Relative molecular mass refers to the mass of a molecule of a compound relative to the mass of an atom of the carbon-12 isotope taken as exactly 12; it is the sum of the relative atomic masses of all the atoms in the molecule.

Relative formula mass is the mass of a unit of the compound as represented by its formula, relative to the mass of carbon-12 atom taken as exactly 12; it is the sum of the relative atomic masses of all the atoms in the formula.

b An empirical formula is the chemical formula of a compound that gives the smallest whole number ratio of atoms of elements in a compound. The molecular formula gives the actual numbers of atoms of the elements in a molecule of a compound.

c A mole is a quantity that contains as many elementary units (e.g. atoms, ions or molecules) as there are atoms in exactly 12 g of the cabon-12 isotope. Molar mass is the mass of a mole of a substance.

····continued in right column ▲

2 a F, F, T, T, F, T

b One mole of oxygen gas contains 2 moles of oxygen atoms. The mass of all atoms of an element is different due to isotopes. One mole of a pure substance would contain 6.022×10^{23} particles of that substance.

3 The steps in calculating the relative molecular mass of a molecule and relative formula mass of an ionic compound are the same. The relative atomic masses for each atom are taken, multiplied by the number of each atom and then added together. The only difference is that ionic compounds may contain subscripts outside of brackets, which means everything inside the bracket must be multiplied by the subscript number.

4

Subject	Formula to use	Calculation
Determine the number of moles of gold in a nugget weighing 0.24 kg.	$n = \dfrac{m}{MM}$	$n = \dfrac{0.24}{197.0}$ $= 0.0012$ mol
Given there are 0.013 moles of sugar ($C_6H_{12}O_6$) in a cup of tea, determine the mass of the sugar cube used.	$m = \dfrac{n}{MM}$	$m = 0.013 \times (6 \times 12.01 + 12 \times 1.008 + 6 \times 16.00)$ $= 2.3$ g
2 moles of an unknown element weighs 53.96 g. Determine the element.	$MM = \dfrac{m}{n}$	$MM = \dfrac{53.96}{2}$ $= 26.98$ Element is aluminium.
A 100 g sample of a compound contains 62.1 g C, 27.7 g O and 10.3 g H. Determine the empirical formula of the compound.	$n = \dfrac{m}{MM}$	$n(C) = \dfrac{62.1}{12.01}$ $= 5.17$ mol $n(O) = \dfrac{27.7}{16.00}$ $= 1.73$ mol $n(H) = \dfrac{10.3}{1.008}$ $= 10.2$ mol $Ratio(C) = \dfrac{5.171}{1.73} = 2.99$ or 3 $Ratio(O) = \dfrac{1.73}{1.73} = 1$ $Ratio(H) = \dfrac{10.2}{1.73} = 5.90$ or 6 Empirical formula is C_3OH_6

5 D

$$\frac{3}{0.75} = 4$$

$$\frac{2}{4} = 0.5$$

6 D

$Fe_2 = 111.7$
$SO_4 = 32.07 + (4 \times 16.00) = 96.07$
$Fe_2(SO_4)_3 = 111.7 + (3 \times 96.07) = 399.9$

7 C

$n(N \text{ atoms}) = 0.28 \times 3 = 0.84$ mol
Mass of N $= 0.84 \times 14.01 = 11.8$ g

8 A

$$n(Na_3PO_4) = \frac{22.8}{163.9} = 0.14 \text{ mol}$$

$$n(O) = 4 \times 0.14 = 0.56 \text{ moles}$$

9 D

Mass H in compound $= n(\text{compound}) \times n(H)$ in compound
$\qquad\qquad\qquad\qquad\qquad \times MM(H)$

A $\dfrac{44}{34.09} \times 2 \times 1.008 = 2.6$ g

B $\dfrac{3}{2.016} \times 2 \times 1.008 = 3$ g

C $\dfrac{13}{58.12} \times 10 \times 1.008 = 2.3$ g

D $\dfrac{22}{16.04} \times 4 \times 1.008 = 5.5$ g

9780170449564

10 a In 100 g sample:

C = 54.5 g, O = 36.4 g, H = 9.1 g

$$n(C) = \frac{54.5}{12.01} = 4.54$$

$$n(O) = \frac{36.4}{16.0} = 2.28$$

$$n(H) = \frac{9.1}{1.008} = 9.03$$

Let $n(O) = 2.28 = 1$

Therefore $n(C) = \frac{4.54}{2.28} = 1.99$ and $n(H) = \frac{9.03}{2.28} = 3.96$

Empirical formula = C_2OH_4

b In 100 g sample:

C = 53.3 g, O = 35.6 g, H = 11.1 g

$$n(C) = \frac{53.3}{12.01} = 4.44$$

$$n(O) = \frac{35.6}{16.00} = 2.23$$

--continued in right column ▲

$$n(H) = \frac{11.1}{1.008} = 11.01$$

Let $n(O) = 2.23 = 1$

Therefore $n(C) = \frac{4.44}{2.23} = 1.99$ and $n(H) = \frac{11.01}{2.23} = 4.94$

Empirical formula = C_2OH_5

$MM(C_2OH_5) = 45.1 \, g\,mol^{-1}$ $MM(compound) = 135 \, g\,mol^{-1}$;

therefore, $\frac{135}{45.1} = 2.99$ times.

Therefore molecular formula = $C_6O_3H_{15}$

c $n(Ag) = 0.412 \times 2 = 0.824$ moles

d $MM(MgSO_4) = 120.38$ $MM(H_2O) = 18.016$

$$n(MgSO_4) = \frac{3.46}{120.38} = 0.0287$$

$n(H_2O) = 2 \times 0.0287 = 0.0574$
$m(H_2O) = 0.0574 \times 18.016 = 1.03 \, g$

11

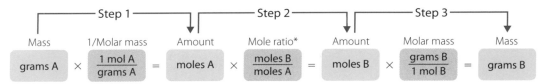

*Mole ratio = ratio of coefficients in balanced chemical equation

---

WS 7.2 PAGE 72

1 1 Weigh a clean crucible and lid.

2 Heat the crucible and lid, allow to cool and reweigh.

3 Repeat step 2 until the crucible and lid have constant mass.

4 Add 5 g of the red cobalt chloride to the crucible with the lid. Reweigh.

5 Heat the crucible on a hot flame, with the lid ajar, until all the cobalt chloride has turned violet in colour.

6 Remove the crucible from the heat and allow to cool.

7 Weigh the crucible with the lid.

8 Return the crucible to the hot flame, with the lid ajar, until all the cobalt chloride has turned blue in colour.

9 Remove the crucible from the heat and allow to cool.

10 Weigh the crucible with the lid.

2 a

Mass of red cobalt chloride (g)	15.81
Mass of violet cobalt chloride (g)	11.06
Mass of water lost (g)	4.75
Mass of blue cobalt chloride (g)	8.66
Mass of water lost (g)	2.40

b $n(CoCl_2) = \frac{8.66}{129.83} = 0.066\,702\,611$ moles $= 6.67 \times 10^{-2}$ moles

c $n(H_2O) = \frac{2.40}{18.016} = 0.133\,214\,92$ moles $= 1.33 \times 10^{-1}$ moles

d ratio $\frac{moles(H_2O)}{moles(CoCl_2)} = \frac{1.33 \times 10^{-1}}{6.67 \times 10^{-2}} = \frac{1.99}{1}$

Therefore ratio is 2:1.
Empirical formula is $CoCl_2.2H_2O$.

e Total mass water lost = 4.75 + 2.40 = 7.15 g.

$$n(H_2O) = \frac{7.15}{18.016} = 0.397 \, mol$$

$n(blue \, CoCl_2) = 6.67 \times 10^{-2} \, mol$

ratio $\frac{moles(H_2O)}{moles(CoCl_2)} = \frac{0.397}{6.67 \times 10^{-2}} = 5.95$

Therefore, the ratio is 6:1.
Empirical formula is $CoCl_2.6H_2O$.

WS 7.3 PAGE 74

1 a $6CO_2(g) + 6H_2O(l) \rightarrow C_6H_{12}O_6(aq) + 6O_2(g)$

b Mole ratio $CO_2 : C_6H_{12}O_6 = 6:1$

Therefore $n(C_6H_{12}O_6) = \frac{1}{6} \times n(CO_2)$

$$= \frac{1}{6} \times 10.0$$

$$= 1.67 \, moles$$

$m(C_6H_{12}O_6) = n \times MM$
$= 1.67 \times 180.156$
$= 301 \, g$

c Mole ratio of $CO_2:H_2O = 6:6$, which equals 1:1, therefore $n(H_2O) = 10.0$ moles.

$$m(H_2O) = n \times MM$$
$$= 10.0 \times 18.016$$
$$= 180\,g, \text{ therefore } 180\,mL$$

2 a $3C(s) + 2H_2O(g) \rightarrow CH_4(g) + 2CO(g)$

b Mole ratio of $C:CH_4 = 2:1$, therefore $n(C) = 2 \times 5.0$
$= 10.0$ moles

$$m(C) = n \times MM$$
$$= 10.0 \times 12.01$$
$$= 120\,g$$

c Mole ratio of $SO_2:H_2SO_4 = 1:1$, therefore $n(SO_2)$
$= 78.0$ moles

$$m(SO_2) = n \times MM$$
$$= 78.0 \times 64.06$$
$$= 499.7\,g$$

d 10.0 kg C produces 0.0070 kg SO_2
$n(SO_2) = 78.0$ moles (as per part **c**)
Mole ratio of $SO_2:H_2SO_4 = 1:1$, therefore $n(H_2SO_4)$
$= 78.0$ moles

Mass C required to produce 78.0 mol $H_2SO_4 = \dfrac{4.99}{0.0070} \times 10.0$
$= 7.1\,kg$

3 a $NaHCO_3(s) + HCl(aq) \rightarrow NaCl\,(aq) + CO_2(g) + H_2O(l)$

b Mass of CO_2 released = 0.91 g.

c The law of conservation of mass states that matter cannot be created nor destroyed, simply transformed; therefore, the mass of the reactants should equal the mass of the products. As the mass of the products is less than the mass of the reactants, it can be assumed that the loss in mass is the result of carbon dioxide gas leaving the liquid products.

d $n(CO_2) = \dfrac{m}{MM} = \dfrac{0.91}{44.01} = 0.0207 = 0.021$ moles

Mole ratio of $NaHCO_3$ to $CO_2 = 1:1$, therefore $n(NaHCO_3)$
$= 0.021$ moles

$m(NaHCO_3) = \dfrac{n}{MM} = 0.021 \times 84.01 = 1.76\,g$

e
$$\%error\,(NaHCO_3) = \frac{\text{theoretical mass} - \text{experimental mass}}{\text{theoretical mass}} \times 100$$
$$= \frac{1.90 - 1.76}{1.90} \times 100$$
$$= \frac{0.14}{1.90} \times 100$$
$$\%error = 7.37\%$$

f

▶ The mass of the carbon dioxide was determined by calculating the difference in mass between products and reactants. Random and systematic errors may occur in the weighing and reading of mass.

▶ There may not have been enough HCl added to react with all the sodium bicarbonate in the tablet; therefore, less carbon dioxide was produced.

▶ Some carbon dioxide may have dissolved into the solution.

▶ The 'fizz' from the reaction may have spattered liquid out of the beaker, reducing the final mass.

WS 7.4 PAGE 77

1 a i Sandwiches made (bread) =
$$\frac{\text{total pieces}}{\text{pieces required per sandwich}} = \frac{300}{2} = 150 \text{ sandwiches}$$

ii Sandwiches made (bacon) =
$$\frac{\text{total pieces}}{\text{pieces required per sandwich}} = \frac{440}{3} = 146 \text{ sandwiches}$$

iii Sandwiches made (eggs) =
$$\frac{\text{total pieces}}{\text{pieces required per sandwich}} = \frac{200}{1} = 200 \text{ sandwiches}$$

b Bacon — as there is only enough to make 146 complete sandwiches.

c Excess (bread) = max. number using bread − max. number using bacon = 150 − 146 = 4 sandwiches worth; therefore, as 2 pieces of bread are used per sandwich, $2 \times 4 = 8$ pieces of bread left.

Excess (egg) = max. number using egg − max number using bacon = 200 − 146 = 54 eggs left (no need to multiply as only 1 egg is used per sandwich).

d Number of sandwiches needed = $\dfrac{1000}{2.5} = 400$

Bread = $400 \times 2 = 800$ pieces of bread
Bacon = $400 \times 3 = 1200$ pieces of bacon
Eggs = $400 \times 1 = 400$ eggs

2 a $Mg(s) + 2HCl(aq) \rightarrow MgCl_2(aq) + H_2(g)$

b Mole ratio $Mg:HCl = 1:2$, therefore $2 \times 2 = 4$ moles.

c Mole ratio $Mg:H_2 = 1:1$, therefore 4 moles.

d i Mole ratio $Mg:HCl = 1:2$, therefore $\dfrac{1}{2} = 0.5$ moles.

ii Mole ratio $Mg:H_2 = 1:1$, therefore 0.5 moles.

e Mole ratio $Mg:HCl = 1:2$, hence 1 mole = $MM(Mg) = 24.31\,g$

f Mole ratio $Mg:H_2 = 1:1$, therefore $MM(Mg):MM(H_2)$

$$\frac{\text{mass}}{MM(Mg)} = \frac{48.62}{24.31} = 2, \text{ therefore } MM(H_2) \times 2 =$$

$2.016 \times 2 = 4.032\,g$

g Mole ratio $Mg:HCl = 1:2$, therefore $n(Mg) =$

$$\frac{n(HCl)}{2} = \frac{0.25}{2} = 0.125 \text{ moles, hence } m(Mg) = n \times MM$$

$= 0.125 \times 24.31 = 3.039 = 3.04\,g$

h i $n(Mg) = \dfrac{m}{MM} = \dfrac{2.0}{24.31} = 0.082$ moles

If 0.082 moles is used, then by the mole ratio $Mg:HCl = 1:2$, then $0.082 \times 2 = 0.164$ moles of hydrochloric acid is required.
If the full 0.2 moles of hydrochloric acid was to be used, then we would require $\dfrac{0.2}{2} = 0.1$ moles = $n \times MM$

$= 0.1 \times 24.31 = 2.4\,g$ of magnesium.
Magnesium is, therefore, the limiting factor.

ii Excess $n(HCl) = n(HCl)$ − required $n(HCl) = 0.2 − 0.164$
$= 0.036$ moles.

3 a $4Fe(s) + 3O_2(g) \rightarrow 2Fe_2O_3(s)$

b Mole ratio $Fe:O_2 = 4:3$, which is 1:0.75, therefore $n(O_2)$
$= 8 \times 0.75 = 6$ moles.

c Mole ratio $Fe:Fe_2O_3 = 4:2$, which is 2:1, therefore $n(Fe_2O_3)$
$= 1$ mole.

d i Mole ratio $Fe:O_2 = 4:3$, therefore $n(Fe) = \dfrac{4}{3} \times 1.5$

$= 2$ moles.

ii Mole ratio $Fe:Fe_2O_3 = 4:2$, which is 2:1, therefore $n(Fe_2O_3)$
$= 1$ mole.

e Mole ratio $Fe:O_2 = 4:3$, therefore $n(Fe) = 4$
$m(Fe) = n \times MM = 4 \times 55.85 = 223.4\,g$

f $n(\text{Fe}) = \dfrac{m}{MM} = \dfrac{111.7}{55.85} = 2$ moles

Mole ratio Fe:Fe$_2$O$_3$= 4:2, which is 2:1, therefore $n(\text{Fe}_2\text{O}_3)$ = 1 moles.
$m(\text{Fe}_2\text{O}_3) = n \times MM = 1 \times 159.7 \times 1 = 159.7\,\text{g}$

g Mole ratio Fe:O$_2$ = 4:3 $n(\text{Fe}) = \dfrac{0.8}{3} \times 4 = 1.07$ moles.

$m(\text{Fe}) = n \times MM = 1.07 \times 55.85 = 59.76\,\text{g}$

h **i** $n(\text{Fe}) = \dfrac{m}{MM} = \dfrac{5.0}{55.85} = 0.0895$ moles.

$n(\text{O}_2)$ required $= \dfrac{3}{4} \times 0.0895 = 0.067$ moles but only

0.0090 moles supplied.

$n(\text{Fe})$ required $= \dfrac{4}{3} \times 0.0090 = 0.012$ moles, therefore

the limiting reagent is the oxygen.

ii Excess $n(\text{Fe}) = n(\text{Fe}) -$ required $n(\text{Fe}) = 0.0895 - 0.012$ $= 0.0775$ moles $= 0.078$ moles.

iii Mole ratio Fe:Fe$_2$O$_3$= 4:2, $n(\text{Fe})$ required = 0.012,

therefore $\dfrac{0.012}{2} = 0.006$ moles.

$m(\text{Fe}_2\text{O}_3) = n \times MM = 0.006 \times 159.7 = 0.96\,\text{g}$

Chapter 8: Concentration and molarity

WS 8.1 PAGE 80

1 Concentration is the amount of solute present in a specified volume of solvent or solution, whereas molarity is the number of moles of solute per litre of solution.

2 a $MM(\text{CH}_4\text{N}_2\text{O}) = 12.01 + (1.008 \times 4) + (14.01 \times 2) + 16.00$ $= 60\,\text{g}$
$m(\text{CH}_4\text{N}_2\text{O}) = n \times MM = 2 \times 60 = 120\,\text{g}$

b 5% (w/v) contains 5 g urea in 100 mL of deionised water.

c 12% (w/w) contains 12 g urea in 100 g of deionised water.
Therefore $m(\text{CH}_4\text{N}_2\text{O}) = 12\,\text{g} \times 2.5 = 30\,\text{g}$

d $n(\text{CH}_4\text{N}_2\text{O}) = \dfrac{m}{MM} = \dfrac{4.5}{60} = 0.075$ mol

$c(\text{CH}_4\text{N}_2\text{O}) = \dfrac{n}{V} = \dfrac{0.075}{0.120} = 0.625\,\text{mol L}^{-1}$

3 a $n(\text{MgCl}_2) = c \times V = 0.0125 \times 0.250 = 0.00313$ moles

b Molar ratio of MgCl$_2$ is 1:2 meaning that MgCl$_2$ contains 1 mole Mg^{2+} and 2 moles Cl$^-$.
Therefore $n(\text{Cl}^-) = 0.00313 \times 2 = 0.00626$

c Molar ratio of MgCl$_2$ is 1:2, meaning that MgCl$_2$ contains 1 mole Mg^{2+} and 2 mole Cl$^-$.
Number of ions $= 0.00313 \times 3 \times 6.02 \times 10^{23} = 5.65 \times 10^{21}$ ions.

d $MM(\text{MgCl}_2) = 95.2$
$m(\text{MgCl}_2) = n \times MM = 0.00313 \times 95.2 = 0.298\,\text{g}$

4 $MM(\text{NaClO}) = 74$
In 1 L, $m(\text{NaClO}) = n \times MM = 0.25 \times 74 = 18.5\,\text{g}$

$m(\text{NaClO})$ in 100 g of solution $= \dfrac{18.5}{1000} \times 100 = 1.85\%$ (w/w).

5 a $\text{Mg(OH)}_2(s) + 2\text{HCl(aq)} \rightarrow \text{MgCl}_2(\text{aq}) + 2\text{H}_2\text{O(l)}$

b $m(\text{Mg(OH)}_2) = \dfrac{3.8 \times 75.0}{100} = 2.85\,\text{g}$

$MM(\text{Mg(OH)}_2) = 58$

$n((\text{Mg(OH)}_2) = \dfrac{m}{MM} = \dfrac{2.85}{58.3} = 0.049$ moles

c Mole ratio Mg(OH)$_2$(s) + 2HCl = 1:2, therefore $n(\text{HCl})$ = 0.049 × 2 = 0.098 moles.

d $V(\text{HCl}) = \dfrac{n}{c} = \dfrac{0.098}{0.30} = 0.33\,\text{L}$

6 a $n(\text{H}_2\text{SO}_4) = c \times V = 14.0 \times 25.0 = 350$ moles
$m(\text{H}_2\text{SO}_4) = n \times MM = 350 \times 98.086 = 34.3\,\text{kg}$

b Mass of solution $= (25.0 \times 1000) \times 1.840 = 46.0\,\text{kg}$

Percentage composition $= \dfrac{34.3}{46.0} \times 100 = 74.6\%$ w/w

c Total $n(\text{H}_2\text{SO}_4) = (c \times V) + (c \times V)$
$= (0.2 \times 0.100)$
$+ (1.2 \times 0.200) = 0.26$ mol
Total volume $(\text{H}_2\text{SO}_4) = 0.100 + 0.200 = 0.300\,\text{L}$

$c(\text{H}_2\text{SO}_4) = \dfrac{n}{V} = \dfrac{0.26}{0.300} = 0.87\,\text{mol L}^{-1}$

WS 8.2 PAGE 83

1

Concentrated solution   Dilute solution

2 a

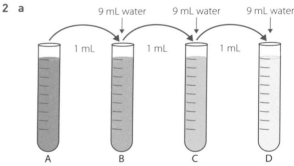

9 mL water   9 mL water   9 mL water

1 mL   1 mL   1 mL

A   B   C   D

10 mol L^{-1} LiCl   1 mol L^{-1} LiCl   0.1 mol L^{-1} LiCl   0.01 mol L^{-1} LiCl

b $c_1V_1 = c_2V_2$; $c_1= 1\,\text{mol L}^{-1}$, $V_1 = ?$, $c_2 = 0.72$, $V_2 = 10\,\text{mL}$

$V_1 = \dfrac{c_2V_2}{c_1} = \dfrac{0.72 \times 10}{1} = 7.2\,\text{mL}$

c While the math calculation is correct, the student's understanding of concentration and dilution is incorrect. The student wants to create a 0.02 mol solution in a 10 mL test tube but required 20 mL (twice as much) of liquid from test tube D to do so. For a dilute solution to become more concentrated, some of the solvent must be removed first, either through evaporation or distillation.

3 a $n(\text{CuCl}_2) = \dfrac{m}{MM} = \dfrac{4.49}{134.45} = 0.0334$ moles

$c(\text{CuCl}_2) = \dfrac{n}{V} = \dfrac{0.334}{0.0515} = 0.649\,\text{mol L}^{-1}$

$c_1V_1 = c_2V_2$; $c_1 = 7.90\,\text{mol L}^{-1}$, $V_1 = 133\,\text{mL}$,
$c_2 = 0.649\,\text{mol L}^{-1}$, $V_2 = ?$

$$V_2 = \frac{c_1 V_1}{c_2} = \frac{7.90 \times 133}{0.649} = 1619\,\text{mL}$$

b Measuring cylinder: $\frac{1619}{100} \times 0.5 = \pm 8.1\,\text{mL}$

Burette: $\frac{1619}{50} \times 0.2 = \pm 6.5\,\text{mL}$

~continued in right column ▲

Beaker: $\frac{1619}{500} \times 1 = \pm 3.2\,\text{mL}$

While the beaker has the largest uncertainty value compared to the other glassware, the uncertainty value is only multiplied once due to the large volume of water the beaker can transfer. Other glassware may have smaller uncertainty values; however, their repeated use increases the potential for error each time.

Chapter 9: Gas laws

WS 9.1 PAGE 85

1

	Solid	Liquid	Gas
Spacing	Close together	Close together	Far apart
Arrangement	Regular pattern	Random arrangement	Random arrangement
Movement of particles	Vibrate on the spot	Move around each other	Move rapidly and freely
Diagram			

2

Water phase change graph

[Temperature °C vs Heat (thermal energy) graph showing SOLID, Melting → / ← Freezing at MP (0°C), LIQUID, Evaporation → / ← Condensation at BP (100°C), GAS]

3

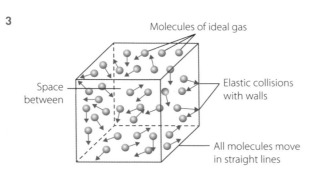

Molecules of ideal gas

Space between — Elastic collisions with walls — All molecules move in straight lines

4 a Individual molecules have differing amounts of kinetic energy.

b $O_2 = 350\,\text{m s}^{-1}$, $N_2 = 400\,\text{m s}^{-1}$, $H_2O = 500\,\text{m s}^{-1}$, $He = 1100\,\text{m s}^{-1}$, $H_2 = 1650\,\text{m s}^{-1}$

c $MM(O_2) = 32\,\text{g}$, $MM(N_2) = 28.02\,\text{g}$, $MM(H_2O) = 18.02\,\text{g}$, $MM(He) = 4.003\,\text{g}$, $MM(H_2) = 2.016\,\text{g}$

d As the molecular/atomic mass of a gas decreases/increases, the average molecular speed increases/decreases.

e The heavier the particles, such as O_2 and N_2, the less variation in speed occurs, represented by narrow curves. The lighter the particles, such as H_2 and He, the broader the curves, indicating a larger variation in speed.

WS 9.2 PAGE 88

1

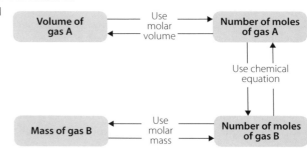

[Flow diagram: Volume of gas A ⇄ (Use molar volume) ⇄ Number of moles of gas A; Number of moles of gas A ⇄ (Use chemical equation) ⇄ Number of moles of gas B; Number of moles of gas B ⇄ (Use molar mass) ⇄ Mass of gas B]

2 This statement is correct. According to Avogadro's law, at standard temperature and pressure, 1 mole of any gas occupies 22.71 L. As CO_2 and O_2 are both gases, they will contain the same number of moles when occupying the same volume. MgO and BaO are solids, so their number of moles is determined by $n = \frac{m}{MM}$ and therefore will not be the same.

3 All balloons are drawn the same size as the mass of each balloon is equal to their molar mass and 1 mole of gas under the same conditions of temperature and pressure occupies the same volume.

4 a The person is exhaling oxygen, carbon dioxide and nitrogen. When these molecules hit the internal walls of the beach ball they cause the pressure within to increase. As more collisions occur, the beach ball becomes larger and firmer until it reaches its maximum volume, or pops.

b If the ball has 1 mole of gas in it, then, on a cold day, the volume of the ball will be 22.71 L and, on a warm day, the volume will be 24.79 L. Therefore, on a cold day, the ball will appear less 'full' and require more breaths to fill completely.

9780170449564

5 a To determine the volume of hydrogen gas produced when a strip of magnesium reacts with dilute sulfuric acid

b

- Magnesium ribbon should be weighed to a similar mass each repetition and recorded.
- The magnesium ribbon should be of similar dimensions, length and width, to ensure that the same surface area of the magnesium ribbon is available to react with the acid.
- Magnesium strip should be cleaned with a piece of steel wool to remove any oxidation to ensure consistent reaction in each repetition.
- The volume of sulfuric acid used in each repetition should be the same, accurately measured with a volumetric flask and recorded.
- The concentration of sulfuric acid used in each repetition should be the same.

c 1 Clean a 5 cm piece of magnesium ribbon with steel wool to remove any oxidation. Weigh the magnesium ribbon. Record the mass in grams.

2 Place the magnesium ribbon into a 1 L beaker containing 600 mL of water.

3 Fill a 50 mL measuring cylinder full to the brim with dilute sulfuric acid and stopper the end.

4 Invert the full measuring cylinder into the beaker and quickly remove the stopper from it making sure to cover the magnesium ribbon.

5 Allow the magnesium ribbon to completely react with the acid. Record the volume of gas produced.

d i $n(Mg) = \dfrac{m}{MM} = \dfrac{2.0}{24.31} = 0.0823$ moles

Molar ratio $n(Mg) : n(H_2) = 1{:}1$, therefore $n(H_2)$ = 0.0823 moles

$V(H_2) = n \times$ molar volume of gas $= 0.0823 \times 24.79 = 2.04$ L

ii The volume of gas exceeds the capacity of measuring cylinder so therefore could not be accurately measured.

e $V(H_2) = 0.0505$ L

$n(H_2) = \dfrac{V(H_2)}{\text{molar volume of gas}} = \dfrac{0.0505}{24.79}$

= 0.002 04 moles

Molar ratio $n(H_2) : n(Mg) = 1{:}1$, therefore $n(Mg)$ = 0.00204 moles

$m(Mg) = n \times MM = 0.002\,04 \times 24.31 = 0.0496$ g

Mass remaining(Mg) = initial mass − final mass

= 2.0 − 0.0496 = 1.95 g

WS 9.3 PAGE 91

1 $V(O_2) = 500 \times 21\% = 105$ L

$V(N_2) = 500 \times 78\% = 390$ L

2 1 atm at the surface + 3.5 atm (35 m depth) = 4.5 atm

1 atm = 101.325 kPa, therefore $101.325 \times 4.5 = 456$ kPa

3 Above sea level, atmosphere pressure very slowly decreases and any change is unnoticeable at 35 m. However, every 10 m below sea level you dive, there is an additional 1 atm of pressure due to the weight of the water above.

4 At the surface, the lungs inhale air at a pressure of 1 atm. At depth, the pressure on the lungs increases and, according to Boyle's law, volume is inversely proportional to the pressure. $P_1V_1 = P_2V_2$ so a diver ascending from 30 m would have a decreased air pressure in the lungs from 4 atm to 1 atm increasing the air volume in the lungs by 4×. If this exceeds the maximum capacity of the lungs, they can rupture. Therefore, by continually exhaling, the expanding air is released, avoiding the possibility of damaging the lungs.

5 1 ft^3 = 28.32 L, therefore 80 ft^3 = 2265.6 L

$P_1 = 1$ atm

$P_1V_1 = P_2V_2$, $P_1 = 1$ atm, $P_2 = 4.5$ atm, $V_1 = 2265.6$ L, $V_2 = ?$

$V_2 = \dfrac{1 \times 2265.6}{4.5} = 503$ L

Each breath is 0.5 L, therefore $\dfrac{503}{0.5} = 1006$ breaths.

6 According to Gay-Lussac's law, as the temperature increases, the pressure increases. A diver filling their tank to 3000 psi may assume it is full; however, when the tank cools, the pressure will decrease. This may leave the diver with less air than they think. Therefore, their time underwater may be less than planned.

7 $\dfrac{P_2}{T_2} = \dfrac{P_1}{T_1}$; $P_1 = 3000$ psi, $P_2 = ?$, $T_1 = 55.0°C + 273 = 328$ K,

$T_2 = 19.0°C + 273 = 292$ K

$P_2 = \dfrac{3000}{328} \times 292 = 2670$ psi

8 Rupture pressure = 3000 + 15% = 3450 psi

$\dfrac{P_2}{T_2} = \dfrac{P_1}{T_1}$; $P_1 = 3000$ psi, $P_2 = 3450$ psi,

$T_1 = 19.0°C + 273 = 292$K, $T_2 = ?$

$T_2 = \dfrac{3000}{292} \times 3450 = 336$ K = 64°C

WS 9.4 PAGE 93

1

Gas law	Graphical representation	Real-life example
Avogadro's law	Volume V vs Moles (n) — straight line through origin, positive slope	As humans inhale, the volume of the lungs increases along with the molar quantity of oxygen.
Gay-Lussac's law	Pressure P vs Temperature T (K) — straight line through origin, positive slope	Pressurised containers have safety labels warning that the container must be kept away from fire and stored in a cool environment or risk explosion.

Gas law	Graphical representation	Real-life example
Boyle's law	Volume V vs Pressure P (decreasing curve)	The 'bends' can occur when a scuba diver ascends to the surface too rapidly. The sudden decrease in pressure causes gases in the blood to expand and may cause organ damage or death.
Charles's law	Volume V vs Temperature T (K) (increasing line)	Helium balloons left out overnight will shrink due to the cold morning air.

2 Real gas has real volume and the collision of the molecules is not elastic because there are intermolecular forces between molecules. As a result, the volume of real gas is much larger than that of the ideal gas, and the pressure of real gas is lower than that of the ideal gas.

3 a $37°C = 37 + 273.15 = 310.15$ K

$$n = \frac{PV}{RT} = \frac{135 \times 25.8}{8.314 \times 310.15} = 1.35 \text{ moles}$$

b $8.20°C = 8.20 + 273.15 = 281.4$ K

$$V = \frac{nRT}{P} = \frac{2.64 \times 8.314 \times 281.4}{233} = 2.65 \text{ L}$$

c $25.8°C = 25.8 + 273.15 = 298.95$ K

$$n = \frac{PV}{RT} = \frac{99.2 \times 12.8}{8.314 \times 298.95} = 0.511 \text{ moles}$$

$m(CO_2) = n \times MM = 0.511 \times 44.01 = 22.5 \text{ g}$

4 a It stays close to 1. It does this because nitrogen gas is behaving like an ideal gas under these conditions.

b As the pressure increases above 200 bar, the compression factor begins to rise. Nitrogen gas is no longer behaving as an ideal gas. As the particles are forced closer, the volume occupied by the molecules themselves relative to the volume of the container becomes significant.

c At 200 K, the compression factor drops significantly and then returns to 1 as the pressure increases up to 200 bar.
At the even colder temperature of 100 K, the compression factor drops dramatically before rising up to 1 again. Again, nitrogen gas is not behaving as an ideal gas. Intermolecular forces become significant as the particles are moving much slower at lower temperatures, having less kinetic energy.

d At 400 K, the compression factor remains close to 1 as the pressure rises up to 200 bar.

e Above 200 bar, the compression factor keeps rising above 1 for all temperatures. At very high pressures, the particles are forced much closer together and the particles themselves occupy much of the available volume.

f The steadily increasing deviation of the compression factor from 1 as the pressure increases above 200 bar becomes less marked as the compression factor is measured at higher temperatures.

g When the gas is at very high pressures or very low temperatures it no longer acts as an ideal gas.

MODULE TWO: CHECKING UNDERSTANDING PAGE 96

1 A-11, B-19, C-5, D-24, E-1, F-16, G-20, H-2, I-25, J-13, K-6, L-3, M-22, N-8, O-9, P-7, Q-4, R-12, S-26, T-10, U-17, V-23, W-15, X-21, Y-14, Z-18

2 A
$PCl_5 + 4H_2O \rightarrow H_3PO_4 + 5HCl$

3 D
$O_2(g) + 2H_2(g) \rightarrow 2H_2O(l)$; Ratio 1:2

4 B
$4.8 + m(O_2) = 17.6 + 14.4$

5 D
$MM(Fe_2(SO_4)_3) = 55.85 \times 2 + (32.07 + 16.00 \times 4) \times 3$

6 C
Percentage of sodium in sample $= \frac{22.99 \times 3}{164} \times 100 = 42\%$

Mass of sample $= 164 \times 0.15 = 24.6 \text{ g}$
Mass of sodium $= 24.6 \times 42\% = 10.3 \text{ g}$

7 B
$V(\text{alcohol}) = v(\text{solution}) \times \%$

8 C
$C_1V_1 = C_2V_2$
$1.50 \times 500 = 0.8 \times V_2 \rightarrow 937.5 - 500 = 437.5 \text{ mL}$

9 B
$P_1V_1 = P_2V_2$
$101.3 \times 3.5 = 92.8 \times V_2 \rightarrow 3.8 \text{ L}$

10 D
$$\frac{4.2 \times 2.1}{0.8} = 11.0 \text{ L}$$

11 C
At high altitude, the gas particles are less likely to make contact with each other as they are further apart. This lack of interaction mimics ideal gas conditions.

MODULE THREE: REACTIVE CHEMISTRY

REVIEWING PRIOR KNOWLEDGE PAGE 98

1 A-17, B-18, C-20, D-3, E-15, F-4, G-12, H-13, I-10, J-11, K-16, L-8, M-14, N-6, O-1, P-7, Q-2, R-9, S-19, T-5

2 a 1 b 2 c 3 d 3 e 0 f 1 g 2 h 1 i 4 j 3 k 0 l 1

3 a $2Mg(s) + O_2(g) \rightarrow 2MgO(s)$
b $CaCO_3(s) \rightarrow CaO(s) + CO_2(g)$
c $CH_4(g) + 2O_2(g) \rightarrow CO_2(g) + 2H_2O(g)$

9780170449564

d $2HNO_3(aq) + Zn(s) \rightarrow Zn(NO_3)_2(aq) + H_2(g)$

e $Na_2CO_3(s) + 2HCl(aq) \rightarrow 2NaCl(aq) + CO_2(g) + 2H_2O(g)$

4 The law of conservation of mass states that in a chemical reaction, matter is neither created nor destroyed. This means the number and types of atom that were present before the reaction are also present after the reaction. The new substances that are formed are just different arrangements of the atoms that were present before the reaction. This means the number and types of atoms present before and after reaction must be the same.

5 a Rate of reaction is a measure of how fast a reaction proceeds.

b Concentration. Increasing concentration increases the rate of reaction.

Temperature. Increasing temperature increases the rate of reaction.

Surface area. Increasing surface area increases the rate of a reaction.

Catalyst. Using a catalyst increases the rate of a reaction.

Chapter 10: Chemical reactions

WS 10.1 PAGE 100

1 a 1-D, 2-C, 3-A, 4-B

b

Equation
$CH_4(g) + O_2(g) \rightarrow CO_2(g) + H_2O(g)$
$CuCO_3(s) \rightarrow CuO(s) + CO_2(g)$
$Fe(s) + S(s) \rightarrow FeS(s)$
$AgNO_3(aq) + NaCl(aq) \rightarrow AgCl(s) + NaNO_3(aq)$

~continued in right column ▲

2 a Precipitation. This is a precipitation reaction because the anions and cations have changed partners and a solid has formed.

b Decomposition. This is a decomposition reaction because the reactant has separated into the two components which were part of the initial compound.

c Synthesis/direct combination. This is a direct combination because the reactants have joined to form a new compound with no additional components.

3 Combustion reaction type is the one missing from question **2**. Combustion reactions can produce a number of different products, so it is difficult to represent this reaction type with a single diagram. For example:

$2Mg(s) + O_2(g) \rightarrow MgO(s)$

$CH_4(g) + 2O_2(g) \rightarrow CO_2(g) + 2H_2O(g)$

The first reaction produces one product (it is also a synthesis reaction), while the second reaction produces two products.

4 a i $2Al_2O_3(s) \rightarrow 4Al(s) + 3O_2(g)$

ii $2H_2O(l) \rightarrow 2H_2(g) + O_2(g)$

b i $2Na(s) + Br_2(g) \rightarrow 2NaBr(s)$

ii $H_2(g) + Cl_2(g) \rightarrow 2HCl(g)$

c i $4Fe(s) + 3O_2(g) \rightarrow 2Fe_2O_3(s)$

ii $C_2H_4(g) + 3O_2(g) \rightarrow 2CO_2(g) + 2H_2O(g)$

d i $3Na_2CO_3(aq) + 2Al(NO_3)_3(aq) \rightarrow Al_2(CO_3)_3(s) + 6NaNO_3(aq)$

ii $Ba(OH)_2(aq) + Cu(NO_3)_2(aq) \rightarrow Cu(OH)_2(s) + Ba(NO_3)_2(aq)$

WS 10.2 PAGE 102

1 a i

	Cl^-	SO_4^{2-}	CO_3^{2-}	OH^-	NO_3^-	O^{2-}
Ca^{2+}	✗	✓	✓	✗	✗	✗
Mg^{2+}	✗	✗	✓	✓	✗	✓
Cu^{2+}	✗	✗	✓	✓	✗	✓
Zn^{2+}	✗	✗	✓	✓	✗	✓
Ag^+	✓ and ✗	✓	✓	✓ and ✗	✗	✓
NH_4^+	✗	✗	✗	✗	✗	✗
Pb^{2+}	✓	✓	✓	✓	✗	✓
Ba^{2+}	✗	✓	✓	✗	✗	✗
Na^+	✗	✗	✗	✗	✗	✗

ii Group 2's result for Ag^+ and Cl^- is different from that of groups 1 and 4. Group 2's result for Ag^+ and OH^- is different from that of group 1. As the results of groups 1 and 4 are consistent, it is possible that group 2 got their results mixed up or used another cation and not Ag^+.

iii Obtain the three silver ion solutions and retest each solution with all the anions to check which results for silver are correct.

b For the ionic compounds tested:

~continued in right column ▲

- all nitrates are soluble
- all chlorides are soluble except for Ag^+ and Pb^{2+}
- all carbonates are insoluble except for Na^+ and NH_4^+
- the solubility/insolubility of compounds, hydroxides and oxides is the same
- all Na^+ and NH_4^+ compounds are soluble.

3 a $Na_2CO_3(aq) + Cu(NO_3)_2(aq) \rightarrow CuCO_3(s) + 2NaNO_3(aq)$

b $Ca(NO_3)_2(aq) + Na_2SO_4(aq) \rightarrow CaSO_4(s) + 2NaNO_3(aq)$

c $AgNO_3(aq) + NaCl(aq) \rightarrow AgCl(s) + NaNO_3(aq)$

4 a $Mg^{2+}(aq) + 2OH^-(aq) \rightarrow Mg(OH)_2(s)$

b $Pb^{2+}(aq) + O^{2-}(aq) \rightarrow PbO(s)$

c $Ba^{2+}(aq) + SO_4^{2-}(aq) \rightarrow BaSO_4(s)$

5 There are many possible tests that can be used. An example is provided.

1 Add a small amount sodium chloride to a small amount of each unknown sample. If a precipitate forms, the sample contains lead.

2 Add a small amount of sodium hydroxide to a small amount of each unknown sample. If a precipitate forms, the sample contains either lead or magnesium. As lead will have already been identified, the sample contains magnesium.

3 The last sample should be barium hydroxide. Use red litmus paper to check for the presence of hydroxide. If paper turns blue, hydroxide is present.

WS 10.3 PAGE 105

1

What are the risks in doing this experiment?	How can you manage risks to stay safe?
Hydrochloric acid is corrosive.	Wear safety glasses and protective clothing. Only use small quantities each time. Take care when handling the chemicals and tell the teacher if there is a spill.
	If hydrochloric acid contacts the skin, immediately wash with lots of water. Tell the teacher.
Limewater is dangerous and can seriously burn the eyes.	Wear safety glasses and do not touch eyes after handling the bottle.
Some chemicals may not be able to be disposed of down the sink.	Refer to safety data sheets (SDS information) or use the RiskAssess program to identify methods of disposal.

2 a The independent variable is the carbonate compounds because these are what will determine the product of the reaction with HCl.

The dependent variable is the products of the reaction as these will be dependent on the what the reactants are.

b The same volume of HCl will be used each time, the temperature of the HCl will be kept the same, the same concentration of HCl will be used for each trial, the mass of carbonate used will be kept the same for each trial, the volume of limewater used will be kept the same, same equipment will be used for each trial.

3 a As it is a qualitative experiment, it is not necessary to control the mass of carbonate, temperature of HCl or volume of limewater. It is important that the equipment is rinsed with distilled water and a fresh sample of limewater is used each time, otherwise there may be contamination causing inaccurate results. The aim states this is an investigation into the reaction between acids and carbonates; however, only one acid is used. Just because carbon dioxide is produced when HCl reacts with different carbonates, it has not confirmed that by using different acids, so has not achieved the aim.

~continued in right column ▲

OR

Although the amount of carbon dioxide being produced is not being measured, there are variables that could have easily been controlled; i.e. mass of carbonate and volume of limewater, which were not controlled. However, controlling these would not have altered the result. The aim states this is an investigation into the reaction between acids and carbonates; however, only one acid is used.

b This investigation does need to be modified to test if the reaction with different acids and carbonates also produces carbon dioxide. The experiment should be repeated using the same volumes and concentration of at least two other acids; e.g. 10 mL of 1 mol L^{-1} HNO_3 and 10 mL of 1 mol L^{-1} H_2SO_4.

OR

This investigation could be modified by specifying the mass of carbonate to be used and also the volume of limewater placed in the test tube. The experiment should be repeated using the same volumes and concentrations of at least two other acids; e.g. 10 mL of 1 mol L^{-1} HNO_3 and 10 mL of 1 mol L^{-1} H_2SO_4.

4 a $2HCl(aq) + CaCO_3(aq) \rightarrow CaCl_2(aq) + CO_2(g) + H_2O(l)$

b acid + carbonate → salt + carbon dioxide + water

5 a Calcium chloride

b Calcium sulfate

c Copper nitrate

Chapter 11: Predicting reactions of metals

WS 11.1 PAGE 108

1 All metals except Pb, Cu, Ag and Au undergo a reaction with air although the rate of the reaction varies from rapidly with K, Na, Li and Ca to slowly with Mg, Al, Zn and Fe. For all the metals except Ag and Au, there is a reaction when heated in oxygen. Pb and Cu react when heated in oxygen, while they had no reaction with air. For all metals that did react in air, the rate of the reaction when heated in oxygen is much more vigorous (e.g. K, Mg) than it was with air at room temperature. This is because for heating in oxygen there is a higher temperature, so reactions will occur more readily and rapidly. There is also a greater concentration of oxygen and an ignition source.

2 a Metals that burn when heated in oxygen should all react to completion: K, Na, Li, Ca, Mg.

e.g. $4K + O_2(g) \rightarrow 2K_2O(s)$

$Mg(s) + O_2(g) \rightarrow 2MgO(s)$

b Aluminium, zinc, copper

e.g. $4Al + 3O_2(g) \rightarrow 2Al_2O_3(s)$

$2Cu(s) + O_2(g) \rightarrow 2CuO(s)$

3

What are the risks in doing this investigation?	How can you manage these risks to stay safe?
Vigorous bubbling may splash material into eyes and onto skin.	Wear safety glasses and protective clothing. Use rice grain size pieces of very reactive metals. Use large beaker of water to absorb heat produced. Cover beaker with gauze mat to prevent material being ejected from the beaker.

▶

9780170449564

What are the risks in doing this investigation?	How can you manage these risks to stay safe?
Reactive metals can seriously burn the skin.	Wear safety glasses and protective clothing. Do not touch reactive metals. Use metal spatula or tweezers to handle the metals. Consult the relevant safety data sheets (SDS information) or use the RiskAssess program to determine risks.
Some chemicals may not be able to be disposed of down the sink.	Refer to SDS information or use the RiskAssess program to identify methods of disposal.

4 a Most reactive (in period 1 and 2): Li, Na, Mg, K, Ca

Least reactive (in period 11): Cu, Ag, Au

b The most reactive metals are on the left-hand side of the periodic table. The least reactive metals are those metals in period 11. Moderately reactive metals are to the right of the periodic table. Generally, reactivity of metals decreases from left to right.

5 a K > Na > Li > Ca > Mg > Al > Zn > Fe > Pb > Cu > Ag > Au

b Aluminium does not appear to react as strongly as expected with steam and acid.

6 a metal + oxygen → metal oxide

b i $4Na(s) + O_2(g) \rightarrow 2Na_2O(s)$

ii $2Ca(s) + O_2(g) \rightarrow 2CaO(s)$

iii $4Fe(s) + 3O_2(g) \rightarrow 2Fe_2O_3(s)$

7 a metal + water → metal hydroxide + hydrogen gas

b i $2K(s) + 2H_2O(l) \rightarrow 2KOH(aq) + H_2(g)$

ii $Mg(s) + 2H_2O(l) \rightarrow Mg(OH)_2(aq) + H_2(g)$

iii Zinc does not react with water at room temperature.

8 a metal + acid → salt + hydrogen gas

b i $2Li(s) + 2HCl(aq) \rightarrow 2LiCl(aq) + H_2(g)$

ii Copper does not react with dilute acids.

iii $Mg(s) + 2HCl(aq) \rightarrow MgCl_2(aq) + H_2(g)$

WS 11.2 PAGE 111

1 a Silver, copper and lead are all heavy metals and are toxic to living organisms, so they should not be poured down the sink where they could pollute the environment. They should be disposed of by proper toxic waste facilities.

b When exposed to air over time most metals react with oxygen to develop an oxide coating. For some metals this oxide coating prevents further reaction. This needs to be removed so the underlying metal can react.

c A metal will not react with itself, so placing a piece of metal in its own solution would be a waste of resources.

2 a $Zn(s) + CuSO_4(aq) \rightarrow Cu(s) + ZnSO_4(aq)$

$Zn \rightarrow Zn^{2+}$ The oxidation number of Zn changes from 0 to +2, so the Zn is oxidised.

$Cu^{2+} \rightarrow Cu$ The oxidation number of copper changes from +2 to 0, so copper is reduced.

b $Fe(s) + Pb(NO_3)_2(aq) \rightarrow Pb(s) + Fe(NO_3)_2(aq)$

$Fe \rightarrow Fe^{2+}$ The oxidation number of Fe changes from 0 to +2, so the Fe is oxidised.

$Pb^{2+} \rightarrow Pb$ The oxidation number of Pb changes from +2 to 0, so Pb is reduced.

3 Based on the number of reactions in the results table, Mg is the most reactive metal as it reacts with all the other metal ions (5 reactions). This means it is the most easily oxidised. Looking at the number of reactions each metal undergoes next would be Al (4 reactions), Zn (3 reactions), Fe (2 reactions) Pb (1 reaction) and lastly Cu does not react with any of the ions of the metals tested; therefore, it is the least reactive metal.

Mg > Al > Zn > Fe > Pb > Cu

4 The results table shows that in all the reactions involving silver, Ag^+ is reduced to Ag, which means Ag^+ oxidises all the metals tested. This means Ag^+ is the most reactive metal ion so Ag would be the least reactive metal. It would be placed after Cu.

Mg > Al > Zn > Fe > Pb > Cu > Ag

5 Al reacts with air to form a protective oxide coating, which prevents further reaction. The most likely error that caused the results obtained is the student did not remove all the oxide coating using the sandpaper before using the pieces of Al metal.

6 a i Pb, Mg, Ag, Al, Zn, Cu, Fe

ii Mg, Al, Zn and Fe are in the same order relative to each other in the metal activity series; however, Pb, Cu and Ag, which are the less active metals, are not at the end. There does not appear to be any clear correlation between atomic radius and metal reactivity for the metals listed.

b i Ag, Cu, Fe, Pb, Zn, Al, Mg

ii Electronegativity describes the power of an atom to attract bonding electrons to it. This means the less active metals will hold on to their electrons (high electronegativity) while the more active metals will be more likely to lose them (low electronegativity). Apart from Fe, the list of electronegativity is the reverse of the metal activity series. There is a strong reverse correlation between electronegativity and metal activity. As one increases the other decreases.

WS 11.3 PAGE 114

1 a True

b False. When copper loses two electrons to form Cu^{2+}, it is oxidised.

c True

d True

e False. For positive monatomic ions, the oxidation state is the charge on the ion.

f False. The oxidation number of manganese in MnO_2 is +4.

g True

h False. In the reaction $CuO(s) + H_2(g) \rightarrow Cu(s) + H_2O(l)$, copper goes from an oxidation state of +2 to 0.

i True

2 a K +1, Br −1

b Mg 0

c Al +3, O −2

d Fe +2, Cl −1

e I 0

f Fe +3, Cl −1

3 a Cl reduced, oxidant; Br oxidised, reductant

b I reduced, oxidant; C oxidised, reductant

c Mn reduced, oxidant; I oxidised, reductant

4 When a metal is a reducer, it is oxidised and electrons are removed from its atoms. The valence electrons are not very strongly held. The removal of its valence electrons results in formation of an ion with an octet of electrons in the outermost shell.

To be an oxidiser, a metal must be reduced by gaining electrons. Metals do not have a strong enough attraction for electrons to take them into their valence shell.

5 $MnO_4^-(aq)$ can oxidise $Cl^-(aq)$ ions but $Cr_2O_7^{2-}(aq)$ cannot; therefore, $MnO_4^-(aq)$ ions are the more powerful oxidant.

6 a Oxidation: $Na(s) \rightarrow Na^+ \text{ (in NaCl)} + e^-$
Reduction: $Cl_2(g) + 2e^- \rightarrow 2Cl^- \text{ (in NaCl)}$
Overall: $2Na(s) + Cl_2(g) \rightarrow 2NaCl(s)$

b Oxidation: $Fe^{2+}(aq) \rightarrow Fe^{3+}(aq) + e^-$
Reduction: $MnO_4^-(aq) + 8H^+(aq) + 5e^- \rightarrow Mn^{2+}(aq) + 4H_2O(l)$
Overall: $5Fe^{2+}(aq) + MnO_4^-(aq) + 8H^+(aq) \rightarrow 5Fe^{3+}(aq) + Mn^{2+}(aq) + 4H_2O(l)$

c Oxidation: $Cu(s) \rightarrow Cu^{2+}(aq) + 2e^-$
Reduction: $NO_3^-(aq) + 2H^+(aq) + e^- \rightarrow NO_2(g) + H_2O(l)$
Overall: $Cu(s) + 2NO_3^-(aq) + 4H^+(aq) \rightarrow Cu^{2+}(aq) + 2NO_2(g) + 2H_2O(l)$

d Oxidation: $2Cl^-(aq) \rightarrow Cl_2(g) + 2e^-$
Reduction: $MnO_2(s) + 4H^+(aq) + 2e^- \rightarrow Mn^{2+}(aq) + 2H_2O(l)$
Overall: $MnO_2(s) + 2Cl^-(aq) + 4H^+(aq) \rightarrow Mn^{2+}(aq) + Cl_2(g) + 2H_2O(l)$

7 a Al is oxidised from 0 to +3.

b Ag is reduced from +1 to 0.

c This is a better cleaning method because no silver is lost from the jewellery as the Ag^+ ions are converted back to Ag metal. Using an abrasive cleaner would remove the Ag_2S layer and silver would be lost.

WS 11.4 PAGE 117

1 Must all the conditions of air, moisture, light and warmth be present for a nail to rust?

2 If a nail is exposed to air and moisture, then it will rust because only these two factors are necessary for rusting to occur.

3 Test tube B

a To check if the nail will rust in the absence of air when water, light and warmth are provided

b All air would have been removed by boiling the water and then, to prevent any further air dissolving in the water, it was covered in a layer of oil.

Test tube C

a This test tube does not have any moisture, so it would be used to check if rusting would occur in the absence of moisture when only air, light and warmth are present.

b By using a desiccant which would absorb all the moisture and sealing the test tube with a bung to prevent further moisture entering

Test tube D

a To check if the nail will rust in the absence of light when moisture, air and warmth are provided

b By wrapping the test tube in foil, light would be prevented from reaching the nail.

Test tube E

a To check if the nail will rust in the absence of warmth when moisture, air and light are provided

b By placing the test tube in the fridge, the temperature would be lowered; however, it would be important to ensure the internal fridge light remains on when the door is closed.

4 a Test tube A would be used to check whether rusting occurs when all four conditions are present.

b The results show that rusting will occur when all four conditions are present.

c It does not prove that all four conditions are necessary for rusting to occur.

5 a

Test tube	A	B	C	D	E
Condition excluded	Nil	Air	Moisture	Light	Warmth
Rusting	Yes	No	No	Yes	Yes

b The results show that rusting does not occur when either air or moisture are absent, so both these conditions are necessary for rusting to occur.

c Neither light nor warmth are necessary for rusting to occur as when light was excluded and the temperature was cooled rusting still occurred.

6 a The investigation did not test freezing temperatures, so no assumption can be made.

b Repeat the investigation but this time use an additional test tube A and place it in the freezer to see how a very low temperature affects the amount of rust produced. As it has already been shown that light is not a condition necessary for rusting, it would not be necessary to include test tube C nor keep a light on in the fridge or freezer.

7 a **i** Fe oxidation number changes from 0 to +3.
ii O oxidation number changes from 0 to −2.

b The change in oxidation number of Fe shows it has been oxidised and the change in oxidation number of O shows it has been reduced.

c $Fe \rightarrow Fe^{3+} + 3e^-$

d Both oxygen and water are necessary for rusting to occur. The formula shows that oxygen has been involved in the reaction as it is present in the formula of the final product, Fe_2O_3. Water is also present as the information provided says the product is hydrated, which means there must be water present. The formula for the final product should have been written, $Fe_2O_3 \cdot xH_2O$, where xH_2O represents the water bound into the final product.

8 $2Fe(s) + O_2(g) + 2H_2O(l) \rightarrow 2Fe(OH)_2(s)$

WS 11.5 PAGE 120

1 a Use the same concentration and volume of electrolyte solutions. Use a new salt bridge each time to prevent contamination. Make sure conditions are kept the same. Calibrate the voltmeter.

b Variables should be controlled to help ensure reliability of results.

2 a To make a salt bridge, which is necessary to complete the circuit

b If a salt bridge wasn't used, the circuit would not be complete and the galvanic cell would not work.

c As reactions occur in the half-cells, the number of positive ions in the solution in the anode half-cell increases as the metal anode is oxidised. The number of positive ions in the cathode half-cell decreases as these are reduced to metal. As both solutions must be electrically neutral, the salt bridge provides a pathway for the movement of anions into the anode half-cell and cations into the cathode half-cell.

3

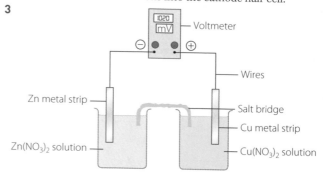

4 A clean salt bridge should be used for each new combination to prevent contamination of the half-cells. If the voltmeter does not produce a reading, the wires will need to be swapped. Electrodes, especially the more reactive metals, react readily in air and form an oxide coating. Each electrode therefore needs to be carefully cleaned with sandpaper to ensure that the metal is exposed to solution in each half-cell.

~continued in right column ▲

5

Beaker 1/ beaker 2	Polarity	Voltage (V)	Anode reaction	Cathode reaction	Theory value	% error
Zinc/copper	±	1.02	$Zn(s) \rightarrow Zn^{2+}(aq) + 2e^-$	$Cu^{2+}(aq) + 2e^- \rightarrow Cu(s)$	1.10	7.3
Zinc/iron	±	0.2	$Zn(s) \rightarrow Zn^{2+}(aq) + 2e^-$	$Fe^{2+}(aq) + 2e^- \rightarrow Fe(s)$	0.32	38
Zinc/lead	±	0.6	$Zn(s) \rightarrow Zn^{2+}(aq) + 2e^-$	$Pb^{2+}(aq) + 2e^- \rightarrow Pb(s)$	0.89	32
Zinc/magnesium	∓	0.9	$Mg(s) \rightarrow Mg^{2+}(aq) + 2e^-$	$Zn^{2+}(aq) + 2e^- \rightarrow Zn(s)$	1.61	44
Iron/copper	±	0.64	$Fe(s) \rightarrow Fe^{2+}(aq) + 2e^-$	$Cu^{2+}(aq) + 2e^- \rightarrow Cu(s)$	0.79	19
Iron/lead	±	0.26	$Fe(s) \rightarrow Fe^{2+}(aq) + 2e^-$	$Pb^{2+}(aq) + 2e^- \rightarrow Pb(s)$	0.32	19
Iron/magnesium	∓	1.41	$Mg(s) \rightarrow Mg^{2+}(aq) + 2e^-$	$Fe^{2+}(aq) + 2e^- \rightarrow Fe(s)$	1.92	27
Lead/magnesium	∓	1.52	$Mg(s) \rightarrow Mg^{2+}(aq) + 2e^-$	$Pb^{2+}(aq) + 2e^- \rightarrow Pb(s)$	2.24	32
Lead/copper	±	0.51	$Pb(s) \rightarrow Pb^{2+}(aq) + 2e^-$	$Cu^{2+}(aq) + 2e^- \rightarrow Cu(s)$	0.47	8.5
Magnesium/copper	±	2.03	$Mg(s) \rightarrow Mg^{2+}(aq) + 2e^-$	$Cu^{2+}(aq) + 2e^- \rightarrow Cu(s)$	2.71	31

6 a i $Zn(s) + Fe^{2+}(aq) \rightarrow Zn^{2+}(aq) + Fe(s)$

ii $Mg(s) + Cu^{2+}(aq) \rightarrow Mg^{2+}(aq) + Cu(s)$

b The combination that produces the lowest voltage is the two metals that are closest together in the table of standard reduction potentials. The combination that produces the highest voltage is the metals that are furthest apart in the table of standard reduction potentials. This means the greater the difference in oxidising (or reducing strength), the larger the voltage produced in the galvanic cell of this combination.

7 According to the results, Mg is the most reactive while Cu is the least reactive. Zn is oxidised by Fe and Pb, so it is more reactive than both these. Fe is oxidised by Pb, which makes it the more reactive metal. The order of ease of being oxidised is: Mg, Zn, Fe, Pb, Cu. This matches the order of these metals in the metal activity series, so there is a direct correlation between metal reactivity and ease of being oxidised.

8 a Values in table

Sample calculations:

Zinc/copper

$Cu^{2+}(aq) + 2e^- \rightarrow Cu(s)$ $E_1^{\ominus} = \varepsilon^{\ominus}_{Cu} = +0.34\ V$

$Zn(s) \rightarrow Zn^{2+}(aq) + 2e^-$ $E_2^{\ominus} = -\varepsilon^{\ominus}_{Zn} = -(-0.76) = +0.76\ V$

$E_{total}^{\ominus} = E_1^{\ominus} + E_2^{\ominus} = 0.34 + 0.76 = 1.10\ V$

Zinc/iron

$Fe^{2+}(aq) + 2e^- \rightarrow Fe(s)$ $E_1^{\ominus} = \varepsilon^{\ominus}_{Fe} = -0.44\ V$

$Zn(s) \rightarrow Zn^{2+}(aq) + 2e^-$ $E_2^{\ominus} = -\varepsilon^{\ominus}_{Zn} = -(-0.76) = +0.76\ V$

$E_{total}^{\ominus} = E_1^{\ominus} + E_2^{\ominus} = -0.44 + 0.76 = 0.32\ V$

b Values in table

Sample calculations:

$$\% \text{ error} = \frac{\text{theoretical error} - \text{experimental value}}{\text{theoretical value}} \times 100\%$$

Zinc/copper

$$\% \text{ error} = \frac{1.10 - 1.02}{1.10} \times 100 = \frac{0.08}{1.10} \times 100 = 7.3\%$$

Zinc/iron

$$\% \text{ error} = \frac{0.32 - 0.2}{0.32} \times 100 = \frac{0.12}{0.32} \times 100 = 38\%$$

c All the experimental values for the voltage are less than the theoretical values except for the lead/copper galvanic cell, which is higher. The order of the experimental values matches the theoretical values except for the zinc/lead cell and the iron/copper cell, which are in the opposite order experimentally to the theoretical values. The values show a range of errors from 7.3% to 44%. The theoretical values are based on standard conditions and it is not stated if these were used. A systematic error, such as an incorrectly calibrated voltmeter, may have contributed to overall lower experimental values. It is difficult to identify specific errors for different cells. Possibly some of the metals may not have been well cleaned to remove the oxide coating, resulting in less reaction. Random errors such as not properly attaching the wires or incorrectly reading the voltmeter may have occurred. Also, as the cells were reused the concentration of the electrolyte solution would have been reduced following each reaction so standard conditions were not used.

Suggestions for improvement would be:

▶ thoroughly clean the metals before each trial

▶ use fresh electrolyte solution for each trial

▶ ensure voltmeter is properly calibrated

▶ ensure wires have clean contact with the electrodes

▶ conduct repeat trials.

9

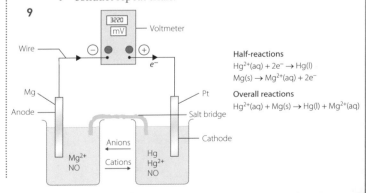

1 a The Ag^+, Ag half-cell is the cathode, and the Cu^{2+}, Cu(s) half-cell is the anode.
Cell voltage $= E^{\ominus}_{cathode} - E^{\ominus}_{anode} = (0.80\,V) - (+0.34\,V)$
$= 0.46\,V$

b The Cu^{2+}, Cu(s) half-cell is the cathode and the $PbSO_4$, Pb half-cell is the anode.
Cell voltage $= E^{\ominus}_{cathode} - E^{\ominus}_{anode} = (+0.34V) - (-0.36V)$
$= 0.70\,V$

c The $PbSO_4$, Pb half-cell is the cathode and the Zn^{2+}(aq), Zn(s) half-cell is the anode.
Cell voltage $= E^{\ominus}_{cathode} - E^{\ominus}_{anode} = (-0.36\,V) - (-0.76\,V)$
$= 0.40\,V$

d The Ag^+, Ag half-cell is the cathode, and the Zn^{2+}, Zn half-cell is the anode.
Cell voltage $= E^{\ominus}_{cathode} - E^{\ominus}_{anode} = (0.80\,V) - (-0.76\,V)$
$= 1.56\,V$
We can see that electrode potentials are additive:
$E^{\ominus}cell\ (d) = E^{\ominus}cell\ (a) + E^{\ominus}cell\ (b) + E^{\ominus}cell\ (c)$

2 This would not change the relative abilities to compete for electrons as these are not dependent on which half-cell is chosen as a reference point. However, the numerical values of the E° would be different.

3 a **i** 5.80 V
ii 5.34 V
iii 4.64 V
iv 4.24 V

b The calculated cell voltages are identical to those calculated in question **1**, demonstrating that the differences of electrode potentials do not depend on the reference point chosen for standard electrode potentials.

4 No, the standard electrode potentials are not absolute measures of competitiveness to attract electrons. Any half-cell could have been selected to use as the reference (and any value of potential attributed to it).

5 a Cell voltage $= \varepsilon^{\ominus}_{Ag} - \varepsilon^{\ominus}_{Sn} = (0.80\,V) - (-0.14\,V) = 0.94\,V$
Spontaneous because the cell voltage is positive.

b Cell voltage $= \varepsilon^{\ominus}_{Ag} - \varepsilon^{\ominus}_{Ni} = (0.80\,V) - (-0.24\,V) = 1.04\,V$
Spontaneous because the cell voltage is positive.

c Cell voltage $= \varepsilon^{\ominus}_{Al} - \varepsilon^{\ominus}_{Fe} = (-1.68\,V) - (-0.44\,V) = -1.24\,V$
Not spontaneous because the cell voltage is negative.

d Cell voltage $= \varepsilon^{\ominus}_{Br} - \varepsilon^{\ominus}_{Fe} = (1.09\,V) - (-0.44\,V) = 1.53\,V$
Spontaneous because the cell voltage is positive.

6 a The strongest reductant is the most easily oxidised while the weakest reductant is the least easily oxidised. From the table of standard reduction potentials:
$Al(s) \to Al^{3+} + 3e^-$ $E^{\ominus} = +1.86\,V$
$Fe(s) \to Fe^{2+} + 2e^-$ $E^{\ominus} = +0.44\,V$
$Pb(s) \to Pb^{2+} + 2e^-$ $E^{\ominus} = +0.13\,V$
$Cu(s) \to Cu^{2+} + 2e^-$ $E^{\ominus} = -0.34\,V$
Therefore, the strongest reductant is Al and weakest is Cu.

b The strongest oxidant is the most easily reduced while the weakest oxidant is the least easily reduced. From the table of standard reduction potentials.
$K^+ + e^- \to K(s)$ $E^{\ominus} = -2.94\,V$
$Al^{3+} + 3e^- \to Al(s)$ $E^{\ominus} = -1.86\,V$
$Ni^{2+} + 2e^- \to Ni(s)$ $E^{\ominus} = -0.24\,V$
$Ag^+ + e^- \to Ag(s)$ $E^{\ominus} = +0.80\,V$
Therefore, the strongest oxidant is Ag^+ and weakest is K^+.

7 a $E^{\ominus}_{total} = E^{\ominus}_{red} + E^{\ominus}_{oxid}$
Cu, Cu^{2+} half-cell is the cathode; $Cu^{2+} + 2e^- \to Cu(s)$
$E^{\ominus} = +0.34\,V$

i $E^{\ominus}_{total} = \varepsilon^{\ominus}_{Cu} + E^{\ominus}_X$
$+0.60 = +0.34 + E^{\ominus}_X$
$E^{\ominus}_X = 0.60 - 0.34 = +0.26\,V$

ii X is oxidised so reduction potential is $-0.26\,V$. Using a table of standard reduction potentials, X is likely to be Ni.
$Ni^{2+} + 2e^- \to Ni(s)$ $E^{\ominus} = -0.26\,V$

iii Possible electrolyte is a soluble Ni^{2+} salt, e.g. $Ni(NO_3)_2$

b $E^{\ominus}_{total} = E^{\ominus}_{red} + E^{\ominus}_{oxid}$
Al, Al^{3+} ion half-cell is the anode (oxidation); $Al(s) \to Al^{3+} + 3e^-$
$E^{\ominus} = +1.68\,V$

i $E^{\ominus}_{total} = E^{\ominus}_Y + \varepsilon^{\ominus}_{Al}$
$+0.50 = E^{\ominus}_Y + 1.68$
$E^{\ominus}_Y = 0.50 - 1.68 = -1.18\,V$

ii Y is reduced so reduction potential is $-1.18\,V$. Using a table of standard reduction potentials, Y is likely to be Mn.
$Mn^{2+} + 2e^- \to Mn(s)$ $E^{\ominus} = -1.18\,V$

iii Possible electrolyte is a soluble Mn^{2+} salt, e.g. $Mn(NO_3)_2$

Chapter 12: Rates of reactions

1 a How does changing the surface area of 2 g of marble chips affect the rate of reaction with excess hydrochloric acid?

b If the surface area of the 2 g of marble chips is increased by making them a powder, then the rate of the reaction with hydrochloric acid will increase because the greater the surface area, the faster the rate of reaction.

c The independent variable is the surface area of the marble chips as this is the variable being used to control the reaction rate. The dependent variable is the rate of production of gas as this rate will be affected by the surface area of the marble chips.

d To ensure valid data, experimental controls need to be in place. The concentration of the HCl must be kept the same as must the mass of marble chips. The same volume of HCl should be used each time. The temperature of the acid and water, apparatus, device to measure time and method should all be kept the same for the two situations.
To ensure reliable data, the investigation should also be replicated by conducting a number of trials. This would provide multiple data sets and allow for checks of precision and reliability.

2 The cylinder was upside down to collect the gas accumulating above the water because the gas is less dense than water. It was filled with water initially to ensure no air was present.

3 Bubbling the gas through limewater would turn the liquid cloudy if the gas was carbon dioxide.
$Ca(OH)_2(aq) + CO_2(g) \to CaCO_3(s) + H_2O(l)$

4 $CaCO_3(s) + 2HCl(aq) \to CaCl_2(aq) + CO_2(g) + H_2O(l)$

5

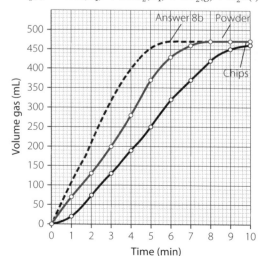

6 Initially, both reactions progressed rapidly as can be seen by the slopes of the graphs between 0 and 5 minutes. In the first 5 minutes, the slopes of the graphs are: $slope_{chips} = \frac{250}{5}$ = 50 mL min^{-1}; $slope_{powder} = \frac{370}{5}$ 74 mL min^{-1}, while between 5 and 10 minutes the slopes become $slope_{chips} = \frac{460 - 250}{5}$ = 42 mL min^{-1}; $slope_{powder} = \frac{470 - 370}{5}$ = 20 mL min^{-1}. Both reactions began to slow down as time progressed and both rates will reach zero as the reaction reaches completion.

7 a The reaction with powdered calcium carbonate had a faster reaction rate. The reaction with powdered calcium carbonate reached completion, as seen by a rate of zero, while the reaction with chips was still progressing.

~continued in right column ▲

b The reaction rate for the powder was much faster than for the chips because of the increased surface area. More calcium carbonate particles were exposed to collisions with acid particles, so more successful collisions forming products occurred.

8 a Using a higher concentration of acid should cause the reaction rate of both chips and powder to increase because, similar to surface area, there are more particles available for collision. However, the powered version should still proceed faster than the chips.

b Answer on graph in question 5.

WS 12.2 PAGE 130

1

Action	Effect on reaction rate	Collision theory explanation
Heating the reactants	Increases	The particles would have more kinetic energy on average and be moving faster. This would result in more frequent collisions and, as more of those collisions would have sufficient energy, more products would be formed.
Using a powdered solid rather than a lump	Increases	An increase in surface area means that more of the solid's particles would be exposed to collisions by dissolved or gaseous particles, which would result in more collisions and therefore more product.
Increasing the pressure on a gas reactant	Increases	Increasing the pressure on a gas reactant in effect increases its concentration per unit volume. Using a more concentrated reactant would result in more frequent collisions between particles and, as a result, more successful collisions.
Cooling a liquid reactant	Decreases	The particles would have less kinetic energy on average and be moving more slowly. This would result in less frequent collisions and, as fewer of those collisions would have sufficient energy, less product would be formed.
Increasing the concentration of a liquid reactant	Increases	Increasing the concentration means there are more particles per unit volume; therefore, more collisions occur, resulting in more product being formed.
Adding more water to dissolved reactants	Decreases	Adding more water effectively reduces the concentration of the dissolved reactants, which results in fewer collisions and less product being formed.
Adding a catalyst to dissolved reactants	Increases	Adding a catalyst reduces the required activation energy and so a greater proportion of collisions will have sufficient energy to initiate the activation complex; therefore, product will be formed more quickly.
Increasing the volume of a gas reactant	Decreases	Increasing the volume reduces the concentration per unit volume, resulting in fewer collisions between particles, and so fewer successful collisions forming products.
Dissolving a solid reactant first	Increases	The reactants will collide much more frequently when both are in a dissolved state.

2 a

	Diagram	Results
Sufficient energy at the correct orientation		
Correct orientation but insufficient energy		
Sufficient energy but wrong orientation		
Insufficient energy and wrong orientation		

9780170449564

ANSWERS 211

b

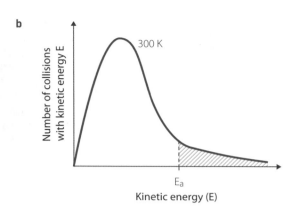

c

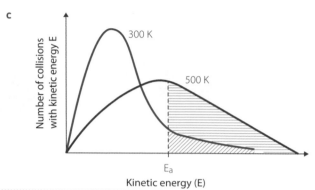

WS 12.3 PAGE 132

1 They should use the same equipment and experimental set-up each time. The measuring cylinder used should be completely filled with water each time and a line marked on it to ensure the same volume of gas is collected each time. For each of the concentration experiments, the length of magnesium ribbon and mass of calcium carbonate used and the temperature and volume of the acid should be kept the same. For the temperature experiments, the concentration and volume of the acid, the length of magnesium ribbon and mass of calcium carbonate used should be kept the same. The magnesium should be cleaned with sandpaper to prevent inactive sites of reaction.

They have decided to conduct repeat trials so this will also help ensure reliability of data.

2

What the risks in doing this investigation?	How can you manage these risks to stay safe?
Chemicals may splash.	Wear safety glasses and wash hands at the end of the experiments.
Hydrogen gas is extremely flammable.	Ensure there are no naked flames and other sources of ignition near the experiment. Work in a well-ventilated area.
Hydrochloric acid is corrosive and could splash onto hands.	Wash hands immediately if they come into contact with the acids. Report any spills. Use the smallest practicable volumes of acid.

3 $2HCl(aq) + Mg(s) \rightarrow MgCl_2(aq) + H_2(g)$
$2HCl(aq) + CaCO_3(s) \rightarrow CaCl_2(aq) + CO_2(g) + H_2O(l)$

~continued in right column ▲

4 a

Concentration of HCl (mol L^{-1})	Time to produce 30 mL of gas (sec)							
	Investigation 1 A (HCl + Mg)				Investigation 2 A (HCl + CaCO$_3$)			
	Trial 1	Trial 2	Trial 3	Average	Trial 1	Trial 2	Trial 3	Average
0.50	306	302	305	304	360	365	363	363
1.0	225	215	224	225	181	184	183	183
1.5	63	61	62	62	118	100	116	117
2.0	22	21	21	21	90	92	91	91

Temperature (°C)	Time to produce 30 mL of gas using 1.5 mol L^{-1} HCl (sec)							
	Investigation 1 B (HCl + Mg)				Investigation 2 B (HCl + CaCO$_3$)			
	Trial 1	Trial 2	Trial 3	Average	Trial 1	Trial 2	Trial 3	Average
25	63	63	65	64	120	122	210	121
35	32	31	32	32	65	65	66	65
45	17	17	17	17	32	30	31	31
55	9	9	10	9	18	17	18	18
65	5	4	4	4	12	11	12	12

b The identified results are anomalous because they are significantly different from the other trial values obtained; therefore, they have been omitted when calculating averages.

5 For both concentration experiments, the gases were produced more quickly as the concentration increased. Therefore, as concentration increases, reaction rate also increases.

6 a

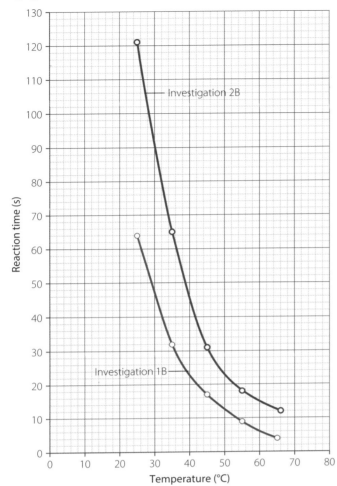

b For both experiments, an increase in temperature resulted in the gas being produced much more quickly; therefore, as the temperature increases the rate of the reaction also increases.

7 a The student was trying to see if changing the temperature and concentration had different effects on reaction rate, depending on the type of reaction being conducted.

b The investigations demonstrate that both temperature and concentration affect the rate at which a reaction occurs. The results demonstrate that for both reactions studied the reaction rate increases with increasing temperature and increasing concentration. Increasing the concentration increases the number of collisions that occur between reactants. Increasing the temperature increases both the rate of collisions and also the energy of collisions. Both reactions show the same effect of concentration and temperature; therefore, the effect of increasing temperature and concentration on reaction rate is not dependent on the reaction being studied.

8 The reaction between HCl and $CaCO_3$ has the higher activation energy, as shown by the large increase in rate between 25°C and 45°C. The slope of the curve between these temperatures is greater than that of the other reaction, which means the increase in temperature has significantly increased the number of particles with sufficient energy to undergo successful reaction.

MODULE THREE: CHECKING UNDERSTANDING PAGE 136

1 A-7, B-13, C-25, D-5, E-2, F-1, G-26, H-23, I-3, J-24, K-17, L-4, M-15, N-12, O-6, P-19, Q-10, R-9, S-18, T-8, U-22, V-11, W-21, X-20, Y-16, Z-14

2 a Sulfur and oxygen or iron and sulfur
$$S(s) + O_2(g) \rightarrow SO_2(g)$$
$$4Fe(s) + 3O_2(g) \rightarrow 2Fe_2O_3(g)$$

b Zinc carbonate
$$ZnCO_3(s) \rightarrow ZnO(s) + CO_2(g)$$

c Hydrochloric acid and magnesium or iron
$$2HCl(aq) + Mg(s) \rightarrow MgCl_2(aq) + H_2(g)$$
$$2HCl(aq) + Fe(s) \rightarrow FeCl_2(aq) + H_2(g)$$

d Calcium hydroxide and sodium carbonate
$$Ca(OH)_2(aq) + Na_2CO_3(aq) \rightarrow MgCO_3(s) + 2NaOH(aq)$$

e Hydrochloric acid and zinc carbonate or sodium carbonate
$$2HCl(aq) + ZnCO_3(s) \rightarrow ZnCl_2(aq) + H_2O(l) + CO_2(g)$$
$$2HCl(aq) + Na_2CO_3(s) \rightarrow 2NaCl(aq) + H_2O(l) + CO_2(g)$$

f Hydrochloric acid and calcium hydroxide
$$2HCl(aq) + Ca(OH)_2(aq) \rightarrow CaCl_2(aq) + 2H_2O(l)$$

3 a Lead(II) chloride, $PbCl_2$

b The new combination of lead(II) and chloride ions would result in a precipitate as lead(II) chloride is insoluble. (The other new combination of aluminium and nitrate ions would not result in a precipitate as all nitrates are soluble.)

c $2AlCl_3(aq) + 3Pb(NO_3)_2(aq) \rightarrow 3PbCl_2(s) + 2Al(NO_3)_3(aq)$

d $2Al^{3+}(aq) + 6Cl^-(aq) + 3Pb^{2+}(aq) + 6NO_3^-(aq) \rightarrow$
$3PbCl_2(s) + 2Al^{3+}(aq) + 6NO_3^-(aq)$

e $Pb^{2+}(aq) + 2Cl^-(aq) \rightarrow PbCl_2(s)$

4 a Highly reactive — Ca, Li
$$Ca(s) + 2H_2O(l) \rightarrow Ca(OH)_2(s) + H_2(g)$$
$$Ca(s) + 2HCl(aq) \rightarrow CaCl_2(aq) + H_2(g)$$
$$2Li(s) + 2H_2O(l) \rightarrow 2LiOH(aq) + H_2(g)$$
$$2Li(s) + 2HCl(aq) \rightarrow 2LiCl(aq) + H_2(g)$$
Moderately reactive — Fe, Zn
$$Fe(s) + 2HCl(aq) \rightarrow FeCl_2(aq) + H_2(g)$$
$$Zn(s) + 2HCl(aq) \rightarrow ZnCl_2(aq) + H_2(g)$$
Not reactive — Cu, Ag

b Place a piece of one metal in a solution containing the metal ion of the other and see which one reacts. The reaction where the metal is displaced from the solution means that that metal is the less reactive of the two because a more reactive metal displaces a less reactive metal from solution.

c Lithium and calcium, which are oxidised by water, are better reducers than zinc, iron, copper and silver. Iron and zinc are oxidised by HCl, so they are better reducers than Cu and Ag, which do not react. On the evidence available, we cannot distinguish between the oxidising abilities of the metals.

5 a I reduced, oxidant
C oxidised, reductant

b I goes from +5 to 0, C goes from +2 to +4

6 a A^{3+} reacts with B so A^{3+} is more easily reduced than B^{2+}.
C_2 doesn't react with A^{2+} so C_2 is less easily reduced than A^{3+}.
C_2 does react with B so C_2 is more easily reduced than B^{2+}.
The order of ease of reduction would be:
$$A^{3+}(aq) + e^- \rightarrow A^{2+}(aq)$$
$$C_2(aq) + 2e^- \rightarrow 2C^-(aq)$$
$$B^{2+}(aq) + 2e^- \rightarrow B(s)$$
Therefore, I would produce a reaction but II and III would not.

b $B^{2+} < C_2 < A^{3+}$

7 a and **b**

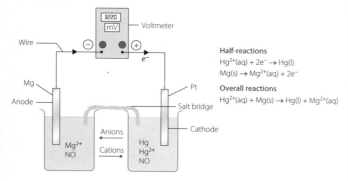

Half-reactions
$Hg^{2+}(aq) + 2e^- \rightarrow Hg(l)$
$Mg(s) \rightarrow Mg^{2+}(aq) + 2e^-$

Overall reactions
$Hg^{2+}(aq) + Mg(s) \rightarrow Hg(l) + Mg^{2+}(aq)$

8 Electrode potentials (measures of competitiveness for electrons) do not depend on the amount of reaction that takes place, so the standard electrode potential for the $Ag^+(aq)$, $Ag(s)$ half-cell should not have been multiplied by a factor of 2.

The correct answer is:

Cell voltage $= E^{\ominus}_{cathode} - E^{\ominus}_{anode} = 0.80\,V - (-0.76\,V) = 1.56\,V$

9 a The steepness of the graph curve indicates how fast a reaction is proceeding. A steep slope (or gradient) indicates that the reaction is fast; a flatter slope indicates the reaction is slow.

b The reaction rate as shown by the gradient of the curve was greatest at the start of the reaction then gradually slowed to zero between 15 s and 80 s.

c The high concentration of the hydrochloric acid at the start of the reaction causes the very fast initial reaction rate, but as the reaction proceeded the acid concentration fell and there was less magnesium, so there were fewer collisions between acid particles and magnesium atoms. The rate of the reaction becomes zero once one of the reactants is exhausted.

10 Reaction B has a steeper slope, so this reaction occurs faster. Reaction B must have occurred at a higher temperature than reaction A. As the temperature increases, more molecules have enough energy to overcome the activation energy barrier so more successful reactions will occur.

MODULE FOUR: DRIVERS OF REACTIONS

REVIEWING PRIOR KNOWLEDGE PAGE 142

1 a i $2C_8H_{18}(l) + 25O_2(g) \rightarrow 16CO_2(g) + 18H_2O(l)$
ii $6CO_2(g) + 6H_2O(l) \rightarrow C_6H_{12}O_6(aq) + 6O_2(g)$
iii $NaOH(s) + HCl(aq) \rightarrow NaCl(aq) + H_2O(l)$

b Bonds broken are: $14 \times C–C$ and $36 \times C–H$ and $25 \times O=O$;
Bonds formed are: $32 \times C=O$ and $36 \times O–H$.

2 Reaction rate can be increased by increasing the temperature, increasing the surface area, increasing the concentration of reactants and by using a catalyst.

3 a A catalyst is a substance that increases the rate of a reaction without undergoing permanent chemical change in the reaction.

b Catalysts usually work by providing a reaction pathway of lower activation energy.

c

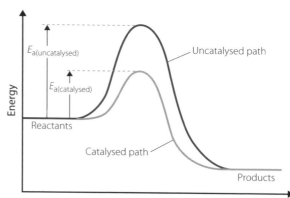

4 a A spontaneous reaction is one that occurs as written. Example: The reaction between zinc metal and copper sulfate to produce copper metal and zinc sulfate is spontaneous, while the reaction between copper metal and zinc sulfate will not occur, so is not spontaneous.

b A reaction with a very high activation energy is less likely to be spontaneous because a lot of energy would need to be added to the system before the reaction would occur. For example, the combustion of a fuel such as propane will only occur if energy such as a flame or a spark is used.

5 a i Polar covalent bonds between H and O.
ii Hydrogen bonds between H on one molecule and O on another.
iii Ionic bonds between Na^+ and Cl^-.

b Ionic substances dissolve because the attractions between the ions and water are stronger than the attractions between oppositely charged ions in the solid.

6 The reaction produces hydrogen gas according to the following equation:

$Mg(s) + 2HCl(aq) \rightarrow MgCl_2(aq) + H_2(g)$

The student, therefore, needs to ensure the reaction occurs in a closed container, otherwise the gas will be released into the atmosphere. If this were to happen, the final mass would be less than the initial mass and the law of conservation of mass would not be demonstrated.

7 a Standard state of a substance is the reference point used to calculate its properties under different conditions, while standard conditions usually refer to temperature of 273.15 K, pressure of 100 kPa and concentration of $1\,mol\,L^{-1}$. Standard conditions are mainly for experimental measurements and allow comparisons to be made between sets of data.

b Standard state for chemical measurements are denoted by the superscript$^{\ominus}$, for example:

Standard electrode potential is represented by $\varepsilon^{\ominus}$ and is the potential of that electrode in its standard state relative to the standard hydrogen electrode.

9780170449564

1 a
1. Measure 100 mL of distilled water into a polystyrene cup.
2. Using a thermometer, measure and record the initial temperature of the water.
3. Accurately weigh 10.0 g of solid aluminium chloride and add this to the same cup.
4. Stir with a stirring rod to ensure the aluminium chloride completely dissolves.
5. Using a thermometer, measure and record the highest, or lowest, temperature reached.
6. Measure 100 mL of distilled water into a second polystyrene cup.
7. Using a thermometer, measure and record the initial temperature of the water.
8. Accurately weigh 5.0 g of solid ammonium nitrate and add this to the second cup.
9. Stir with a stirring rod to ensure the ammonium nitrate completely dissolves.
10. Using a thermometer, measure and record the highest, or lowest, temperature reached.

b

	$AlCl_3$	NH_4NO_3
Mass of solid (g)	10.0	4.9
Initial temperature (°C)	20.0	20.1
Final temperature (°C)	46.3	16.8

c

$$q(AlCl_3) = mC\Delta T$$
$$= 100 \times 4.18 \times (46.3 - 20.0)$$
$$= 10993.4\,J = +10.99\,kJ$$

$$n(AlCl_3) = \frac{10.0}{26.98 + 106.35}$$
$$= 0.075\,mol$$

$$\Delta H = \frac{q}{n}$$
$$= \frac{10.99}{0.075}$$
$$= 146.5\,kJ\,mol^{-1}$$

$$q(NH_4NO_3) = mC\Delta T$$
$$= 100 \times 4.18 \times (16.8 - 20.1)$$
$$= -1379.4\,J\ or\ -1.38\,kJ$$

$$n(NH_4NO_3) = \frac{4.9}{28.02 + 4.04 + 48}$$
$$= 0.061\,mol$$

$$\Delta H = \frac{q}{n}$$
$$= \frac{-1.38}{0.061}$$
$$= -22.6\,kJ\,mol^{-1}$$

d Energy is lost or gained by surroundings.
The solid may not have completely dissolved.

The polystyrene cup may have absorbed a small amount of energy.

Highest or lowest temperature may not have been accurately judged.

Scales may not have been calibrated before use, resulting in inaccurate masses being used.

e $AlCl_3(s) \rightarrow Al^{3+}(aq) + 3Cl^-(aq)$ $\Delta H = +146.5\,kJ\,mol^{-1}$
Reaction is endothermic.
$NH_4NO_3(s) \rightarrow NH_4^+(aq) + NO_3^-(aq)$ $\Delta H = -22.6\,kJ\,mol^{-1}$
Reaction is exothermic.

2 a

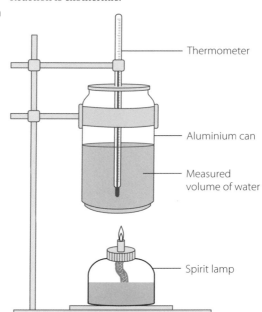

Thermometer

Aluminium can

Measured volume of water

Spirit lamp

b While insulating, a polystyrene cup will melt with the heat from the flame. An aluminium can will conduct heat more efficiently than glass, meaning less heat will be lost to surrounds.

c Methanol: $2CH_3OH(l) + 3O_2(g) \rightarrow 2CO_2(g) + 4H_2O(g)$
Propanol: $2C_3H_7OH(l) + 9O_2(g) \rightarrow 6CO_2(g) + 8H_2O(g)$
Octanol: $C_8H_{17}OH(l) + 12O_2(g) \rightarrow 8CO_2(g) + 9H_2O(g)$

d i Alkanols burned (methanol, propanol and octanol)
ii Change in temperature of the water (°C)
iii
- Volume of water (200 mL)
- Distance between flame and bottom of the can (tip of flame should just touch the can)
- Thermometer height secured so to not touch the bottom of the can
- 350 mL aluminium can

e

What are the risks in doing this investigation?	How can you manage these risks to stay safe?
All alkanols are highly flammable.	Spirit burners are used to contain the fuel and separate them from the flame. Use in a well-ventilated area.
Glassware and other equipment will become hot and may cause burns.	Use beaker tongs or similar to handle hot glassware. Allow equipment time to cool before packing away.

f

Alkanol	Methanol (CH_3OH)	Propanol (C_3H_7OH)	Octanol ($C_8H_{17}OH$)
Mass of water (g)	200	200	200
Initial temperature of water (°C)	17.2	17.5	17.3
Final temperature of water (°C)	42.5	59.8	80.6
ΔT	+25.3	+42.3	+63.3
$q = mc\Delta T$ (J)	21 150.8	35 362.8	52 918.8
Mass of fuel burnt (g)	2.2	1.8	2
ΔH (kJ mol^{-1})	−306.5	−1182.61	−3436.36

i $q(CH_3OH) = mC\Delta T$
$$= 200 \times 4.18 \times (42.4 - 17.2)$$
$$= 21150.8\,J$$

$$n(CH_3OH) = \frac{2.2}{12.01 + 4.032 + 16.00}$$
$$= 0.069\,mol$$

$$\Delta H(CH_3OH) = \frac{-q}{n}$$
$$= \frac{-21.15}{0.069}$$
$$= -306.5\,kJ\,mol^{-1}$$

ii $q(C_3H_7OH) = mC\Delta T$
$$= 200 \times 4.18 \times (59.8 - 17.5)$$
$$= 35362.8\,J$$

$$n(CH_3OH) = \frac{1.8}{36.03 + 8.064 + 16.00}$$
$$= 0.0299\,mol$$

$$\Delta H(CH_3OH) = \frac{-q}{n}$$
$$= \frac{-35.36}{0.0299}$$
$$= -1182.6\,kJ\,mol^{-1}$$

Actual energy $= \dfrac{35.36 \times 100}{75} = 47.15\,kJ$

iii $q(C_8H_{17}OH) = mC\Delta T$
$$= 200 \times 4.18 \times (80.9 - 17.3)$$
$$= 52918.8\,J$$

$$n(C_8H_{17}OH) = \frac{2.0}{96.08 + 18.144 + 16.00}$$
$$= 0.0154\,mol$$

$$\Delta H(C_8H_{17}OH) = \frac{-q}{n}$$
$$= \frac{-52.92}{0.0154}$$
$$= -3436.36\,kJ\,mol^{-1}$$

iv Heat is often lost to surroundings in school experimental procedures. It is also assumed that complete combustion of the fuel occurred when it is more likely that some fuel underwent incomplete combustion, as seen by the build-up of soot on the bottom of the can. The heat may not have been evenly distributed in the can, or the thermometer may have had contact with the can.

WS 13.2 PAGE 149

1 **a** and **b**

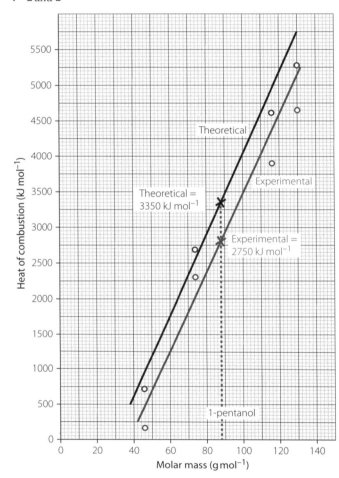

c ▶ The incomplete combustion of the fuel.
  ▶ Heat lost to the surrounding air.
  ▶ Heat loss in heating the conical flask as it is made of glass.

d ▶ Burn fuel in a well-ventilated air to ensure complete combustion of the fuel.
  ▶ Ensure the flame 'licks' the bottom of the flask to minimise heat loss to surrounds. Insulating sides could be put in place if they do not compromise oxygen availability.
  ▶ Replace conical flask with an aluminium container as it is a better conductor of heat.

9780170449564

Alcohol	Molar mass (g mol^{-1})	Experimental heat of combustion (kJ mol^{-1})	Theoretical heat of combustion (kJ mol^{-1})	Experimental heat of combustion (kJ g^{-1})
Ethanol	46	160	720	3.5
1-butanol	74	2300	2670	31
1-heptanol	116	3880	4630	33.5
1-octanol	130	4690	5285	36.1

f Fuels are rarely pure substances. For example, petrol contains hydrocarbons other than octane, which make it difficult to determine the molar heat of combustion. When comparing fuels, it is easier for consumers to compare the volume they are burning, such as how far a single tank of petrol gets them, rather than per mole, which takes into account the size of the molecule.

2

Advantages	Disadvantages
Ethanol is more likely to undergo complete combustion compared to octane as it only required 3 moles of oxygen per mole burned compared to 12.5 moles of oxygen per mole of octane. Octane: $2C_8H_{18} + 25 O_2 \rightarrow 16CO_2 + 18H_2O$ Ethanol: $C_2H_5OH + 3 O_2 \rightarrow 2CO_2 + 3H_2O$ Ethanol releases 2 moles of carbon dioxide compared to 8 moles per mole of octane, therefore contributing less to the enhanced greenhouse effect. Ethanol blends are up to 30 cents per litre cheaper at the petrol station than non-blended fuels. Ethanol is from a renewable source, whereas octane which comes from fossil fuels.	Ethanol produces less energy per mole, and per gram (720 kJ mol^{-1}, 15.7 kJ g^{-1}) compared to 5471 kJ mol^{-1}, 48.0 kJ g^{-1} of octane), meaning a car will not travel as far per tank compared to octane. Current engines can only support 10% ethanol blends. Higher blends would require engine modification. Farming of crops to produce ethanol will use a lot of water and take away from food crops in already drought-stricken farming areas of Australia.

Although ethanol is a more environmentally friendly fuel to combust, burning under complete conditions and producing less carbon dioxide emissions compared to octane, the overall cost to the consumer means ethanol remains an unsuitable alternative. Consumers would need to purchase new engines to support higher blended fuels, and while ethanol blends may be cheaper to purchase, a consumer requires almost three times the mass compared to octane meaning they will need to fill up their cars more often, mitigating the money saved at the petrol station.

WS 13.3 PAGE 152

1 Your friend at the party represents the catalyst, and the conversation between you and another person is representative of the chemical reaction. When your friend introduces you to someone with common interests, it is referring to the action

~continued in right column ▲

of a catalyst in ensuring correct orientation and energy by lowering activation energy to ensure more successful reactions. Your friend moves on afterwards, which is in reference to a catalyst being unchanged by the chemical reaction taking place. When you are at the party alone and beginning your own conversations, this is an analogy for the slow progression of a reaction with high activation energy and a high likelihood of incorrect orientation for a successful reaction.

2 a The iron catalyst is heterogeneous as it is a solid catalyst providing a surface for reactants in a gaseous state.

b

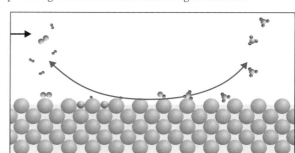

3 a Lactase acts like a pair of scissors, cutting the lactose molecule into the smaller glucose and galactose to be absorbed by the body. When the lactase is not present in sufficient quantity, the lactose cannot undergo decomposition in time to be absorbed by the body. The lactose then reacts with bacteria to produce the symptoms experienced by lactose-intolerant people. The lactase is acting as a catalyst to quickly 'cut' the lactose molecule.

b No. The human body functions in a narrow temperature range. While increasing body temperature can increase the rate of reaction, the increase necessary to decompose the lactose fast enough to be absorbed would be too high to ensure the continued function of other bodily systems.

c

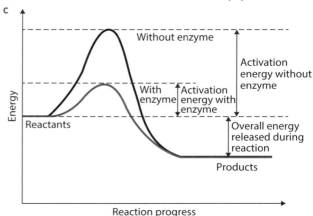

4 a

b If an inhibitor is present, the enzyme cannot act as a catalyst. Without a catalyst present, the activation energy would be greater and the rate of reaction would decrease.

Chapter 14: Enthalpy and Hess's law

WS 14.1 PAGE 155

1 a Adding reactions **i** and **ii** gives the first reaction.
Therefore, $\Delta H = \Delta H_1 + \Delta H_2$
$\Delta H = b + c \, \text{kJ mol}^{-1}$

b

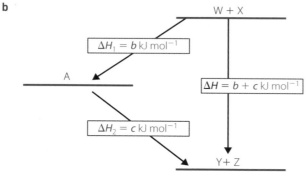

2 i $2 \times (\frac{1}{2} N_2 + \frac{1}{2} O_2 \rightarrow NO(g))$ $\Delta H = 2 \times (+90 \, \text{kJ mol}^{-1})$

ii $2NO(g) + O_2 \rightarrow 2NO_2(g)$ $\Delta H = -112 \, \text{kJ mol}^{-1}$
Adding the above two equations:
$\Delta H = (+180) - 112 \, \text{kJ mol}^{-1} = +68 \, \text{kJ mol}^{-1}$

3 To obtain the first equation for the combustion of two moles of ethane, the following equations need to be added.

i $4 \times (C(s) + O_2(g) \rightarrow CO_2(g))$ $\Delta H = 4 \times (+s) \, \text{kJ}$

ii $6 \times (H_2(g) + \frac{1}{2} O_2(g) \rightarrow H_2O(l))$ $\Delta H = 6 \times (+t) \, \text{kJ}$

iii $2 \times (2C(s) + 3H_2(g) \rightarrow C_2H_6(g))$ $\Delta H = 2 \times (-u) \, \text{kJ}$
$\Delta H = 4s + 6t - 2u \, \text{kJ}$ for the combustion of two moles of ethane. However, enthalpy of combustion of ethane needs to be calculated per mole; therefore, it is necessary to divide the answer by 2, to obtain $\Delta H = 2s + 3t - u \, \text{kJ mol}^{-1}$.

4 i $C(s) + O_2(g) \rightarrow CO_2(g)$ $\Delta H = +s \, \text{kJ}$

ii $2 \times (H_2(g) + \frac{1}{2} O_2(g) \rightarrow H_2O(l))$ $\Delta H = 2 \times (+t) \, \text{kJ}$

iii $CO_2(g) + 2H_2O(l) \rightarrow CH_4(g) + 2O_2(g)$ $\Delta H = (-v) \, \text{kJ}$
$\Delta H = s + 2t - v \, \text{kJ}$ for the heat of formation of methane, CH_4.

WS 14.2 PAGE 157

1 a $H_2(g) + \frac{1}{2} O_2(g)$

b

Path	Enthalpy (kJ mol⁻¹)
1	$\frac{+572}{2} = +286$
2	$\frac{-484}{2} = -242$
3	$+44$

c Addition of enthalpies for paths 1 and 2 gives the same value as for the enthalpy for path 3, as shown below. Therefore, the law of conservation of energy is demonstrated.
ΔH_1 Path 1 + ΔH_2 Path 2 = ΔH_3 Path 3
$+286 \, \text{kJ mol}^{-1} + (-242) \, \text{kJ mol}^{-1} = +44 \, \text{kJ mol}^{-1}$

2 a

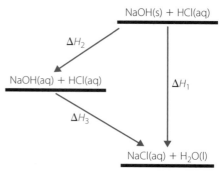

b The enthalpy values obtained by the student support the law of conservation of energy for the reaction of solid sodium hydroxide versus aqueous sodium hydroxide with hydrochloric acid. The enthalpy values obtained in both pathways are the same, $-52.1 \, \text{kJ mol}^{-1}$.

The values for enthalpy obtained from reacting solid sodium hydroxide with hydrochloric acid are:
$NaOH(s) \rightarrow NaOH(aq)$ $\Delta H = -23.2 \, \text{kJ mol}^{-1}$
$NaOH(aq) + HCl(aq) \rightarrow NaCl(aq) + H_2O(l)$
$\Delta H = -28.9 \, \text{kJ mol}^{-1}$

Net enthalpy, $\Delta H = -23.2 \, \text{kJ mol}^{-1} + -28.9 \, \text{kJ mol}^{-1}$
$= -52.1 \, \text{kJ mol}^{-1}$

The enthalpy value for the reaction of solid sodium hydroxide with hydrochloric acid is:
$NaOH(s) + HCl(aq) \rightarrow NaCl(aq) + H_2O(l)$
$\Delta H = -52.1 \, \text{kJ mol}^{-1}$

c i percentage error $= \frac{|58.2 - 52.1|}{58.1} \times 100 = 10.5\%$ (to 3 sig fig)

ii The student's result for enthalpy of neutralisation of aqueous sodium hydroxide with hydrochloric acid is $-52.1 \, \text{kJ mol}^{-1}$, which is less than the theoretical value of $-58.2 \, \text{kJ mol}^{-1}$. Possible reasons for the difference could be due to heat loss to the environment from the foam cup, the concentrations of the sodium hydroxide and hydrochloric acid may not be exactly what was written on the reagent bottles, the volumes measured in the measuring cylinder may not have been accurate and the thermometer may not have been accurate.

iii Possible improvements to minimise errors would be to insulate the foam cup, and use accurate glassware, such as a burette, to deliver exact volumes of sodium hydroxide and hydrochloric acid. The concentration of sodium hydroxide can be determined accurately by titration so that its concentration can be determined accurately for the calculation of enthalpy.

WS 14.3 PAGE 159

1

Experiment 1	Experiment 2
• 2 × 100 mL measuring cylinders • Plastic pipette • Polystyrene cups • 10–110°C thermometer	• 100 mL measuring cylinder • Plastic pipette • Polystyrene cups • −10–110°C thermometer • Top loading balance (to 3 decimal places) • Spatula • 50 mL beaker

2 a Experiment 1: $HCl(aq) + KOH(aq) \rightarrow KCl(aq) + H_2O(l)$

b Experiment 2: $HCl(aq) + KOH(s) \rightarrow KCl(aq) + H_2O(l)$

3

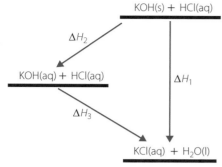

$$KOH(s) + HCl(aq)$$

ΔH_2

$$KOH(aq) + HCl(aq)$$

ΔH_3

ΔH_1

$$KCl(aq) + H_2O(l)$$

4 a $n(H_2O) = n(HCl) = n(KOH) = 0.200 \times 0.054 = 0.0108$

$\Delta T = 21.36°C - 20.00°C = 1.36°C$

$q = mc\Delta T = (54+54) \times 4.18 \times 1.36 = 613.9584\,J$

$\Delta H = \dfrac{q}{n} = \dfrac{613.9584}{0.0108} = 56\,848\,J = -57\,kJ\,mol^{-1}$

b $n(KOH) = \dfrac{0.6060}{(39.10+16.00+1.008)} = -0.01080$

$n(HCl) = 0.200 \times 0.054 = 0.0108$

$n(H_2O) = n(HCl) = n(KOH)$

$\Delta T = 23.72°C - 20.00°C = 3.72°C$

$q = mc\Delta T = 54 \times 4.18 \times 3.72 = -839.6784\,J$

$\Delta H = \dfrac{q}{n} = \dfrac{-839.6784}{0.0108} = -77\,748\,J = -78\,kJ\,mol^{-1}$ (to 2 sig fig)

5 The law of conservation of energy states that the energy change in going from reactants to products is the same regardless of the path taken. The heat of reaction for experiment 1 was $57\,kJ\,mol^{-1}$, while in the second experiment it was slightly higher at $78\,kJ\,mol^{-1}$. The heat of solution is slightly higher in the second experiment because base, KOH, was in a solid state and the heat of solution of KOH accounted for the slightly higher heat of solution compared to when both reactants were in an aqueous form.

6 a Although potassium hydroxide is the limiting reagent, there will be no effect on the heat of reaction because the molar heat of neutralisation will always be $57\,kJ\,mol^{-1}$ for a strong acid and strong base. This is because the reaction is always between hydrogen ions and hydroxide ions to produce water.

$H^+(aq) + OH^-(aq) \rightarrow H_2O(l)$

b $n(HCl) = 0.200 \times 0.054 = 0.0108$

$n(KOH) = 0.100 \times 0.054 = 0.0054 -$ limiting reagent

$n(H_2O) = 0.0054$

$\Delta T = 20.68°C - 20.00°C = 0.68°C$

$q = mc\Delta T = (54 + 54) \times 4.18 \times 0.68 = 306.9792\,J$

$\Delta H = \dfrac{q}{n} = \dfrac{306.9792}{0.0054} = -56\,848\,J = -57\,kJ\,mol^{-1}$

The molar heat of reaction is the same despite a lower concentration of KOH. Therefore, my answer to part **a** above is supported.

WS 14.4 PAGE 162

1 a

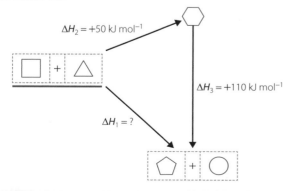

$\Delta H_2 = +50\,kJ\,mol^{-1}$

$\Delta H_3 = +110\,kJ\,mol^{-1}$

$\Delta H_1 = ?$

b $\Delta H_1 = \Delta H_2 + \Delta H_3$

$\Delta H_1 = +50\,kJ\,mol^{-1} + -110\,kJ\,mol^{-1} = -60\,kJ\,mol^{-1}$

c Exothermic, because enthalpy of reaction is a negative value

d

Exothermic reaction

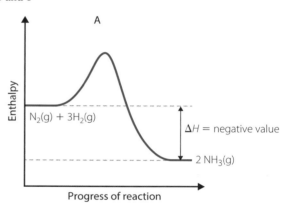

2 a i Matches B

ii Matches A

b and **c**

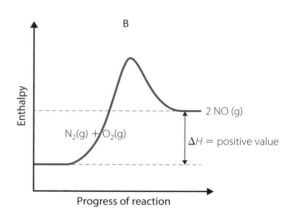

d Energy is required to break bonds, while energy is released when new bonds form. Since the production of NO is endothermic, more energy is required to break the triple covalent bond in N_2 and the double covalent bond in O_2 than the energy released when covalent bonds form in two moles of NO. In contrast, the production of NH_3 is exothermic. Therefore, it can be concluded that the energy required to break the triple covalent bond in N_2 and the three single covalent bonds in three moles of H_2 is less than the energy released when covalent bonds form in two moles of NH_3.

e i $\Delta H^\ominus = \{$sum of standard enthalpies of formation of product compounds$\} - \{$sum of standard enthalpies of formation of reactant compounds$\}$

$\Delta H = (+90 \times 2) - 0 = +180\,kJ\,mol^{-1}$

ii $\Delta H^\ominus = \{$sum of standard enthalpies of formation of product compounds$\}$
$-\ \{$sum of standard enthalpies of formation of reactant compounds$\}$

$\Delta H = (-46 \times 2) - 0 = -92\,\text{kJ}\,\text{mol}^{-1}$

Chapter 15: Entropy and Gibbs free energy

WS 15.1 PAGE 165

1 a Molten wax would have a higher entropy because liquids have a higher entropy than solids of the same substance.

b $Br_2(g)$ would have a higher entropy because gases have a higher entropy than liquids of the same substance.

c Octane would have a higher entropy because it is a larger, more complex molecule.

d Silver chloride would have a higher entropy because it is a more complex molecule.

2 a Decrease, because the number of gas molecules decreases from reactants to product.

b Decrease, because one of the reactants is a gas while the product is only solid.

c Increase, because a gas is formed as a product while the reactant is a solid.

3 a Student A's presentation shows the particles moving to occupy a greater volume and resulting in a decrease in concentration. The particles spread out randomly in all directions and can be found in a greater number of different locations. This therefore increases the entropy of the system.

Student B's presentation demonstrates an increase in entropy by showing the change from the highly ordered crystalline NaCl solid to having Na^+ and Cl^- ions in solution. The ions are more spread out and mixed up rather than being in fixed positions as they are in the crystalline solid. This means the entropy of the system has increased.

Student C's presentation shows two highly ordered crystalline solids (barium hydroxide and ammonium chloride) reacting to form a liquid. The result of the reaction is a liquid, which is much more random than the two original solids, so entropy has increased.

b All three students should have provided information about whether the temperature of the system changed or remained constant, so, determining whether the process was exothermic or endothermic. This would have assisted in determining the drive direction of the energy on the process.

Student C could have conducted research and provided the balanced equation or included information about a gas being produced. The equation for the reaction is:

$Ba(OH)_2.8H_2O(s) + 2NH_4Cl(s) \rightarrow 2NH_3(g) + 10H_2O(l) + BaCl_2(s)$

The presence of the gas cannot be seen in the presentation.

4 a

a In a solid, the particles are close together and just vibrate about fixed positions. It is an orderly arrangement.

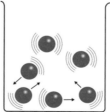

b In a liquid, the particles are further apart and they move from place to place as well as vibrating. This is a less orderly or more chaotic state than in a solid.

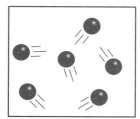

c In a gas, the particles are much further apart and move much more rapidly. This is the least orderly or most chaotic state.

b Absolute zero or 0 K.

c As the temperature increases, the particles absorb more energy and move about more — they have greater kinetic energy. The particles collide and exchange energy with each other, creating more randomness.

d From the graph provided, it can be seen that the entropy of the gaseous state is much greater than the entropy of the liquid state. For example:

Entropy of liquid	Entropy of gas
$H_2O(l)$ $S^\ominus = 70\,\text{J}\,\text{K}^{-1}\text{mol}^{-1}$	$H_2O(g)$ $S^\ominus = 189\,\text{J}\,\text{K}^{-1}\text{mol}^{-1}$
$Br_2(l)$ $S^\ominus = 152\,\text{J}\,\text{K}^{-1}\text{mol}^{-1}$	$Br_2(g)$ $S^\ominus = 245\,\text{J}\,\text{K}^{-1}\text{mol}^{-1}$

The entropy of $H_2O(g)$ is almost 3 times the entropy of $H_2O(l)$, while the entropy of $Br_2(g)$ is over 1.5 times the entropy of $Br_2(l)$. However, the difference in entropies of the two gases is smaller than both these values, so the data supports the statement for this example.

5 a Diamond has a rigid crystalline with strong covalent bonds extending throughout the crystalline lattice. Graphite has strong covalent bonds within each layer but weaker bonds between layers, making it less rigid and more disordered than diamond. Therefore, graphite has higher entropy.

b i As molecules become more complex within a homologous series, their entropy increases. For example, the entropy of methane is 186, while the entropy of butane, a more complex molecule, is 310. The entropy of methanol is 127, while the entropy of the more complex molecule butanol is 228.

ii Molecules that can exhibit hydrogen bonding have lower entropy than molecules of a similar size with no hydrogen bonding. For example, ethanol, which has hydrogen bonding, and ethane, which does not have hydrogen bonding, are similar sized molecules but ethanol has a much lower entropy of 160 compared to ethane's 229.5.

WS 15.2 PAGE 170

1 Consider the following phase change at 298 K:

$H_2O(l) \rightarrow H_2O(g)$

Using data tables:

$\Delta H^\ominus = \Delta H_f^\ominus(H_2O(g)) - \Delta H_f^\ominus(H_2O(l)) = -242 - (-286)$
$= +44\,\text{kJ}\,\text{mol}^{-1}$

$\Delta S^\ominus = S^\ominus(H_2O(g)) - S^\ominus(H_2O(l)) = 189 - 70 = +119\,\text{J}\,\text{K}^{-1}\text{mol}^{-1}$
$= 0.119\,\text{kJ}\,\text{K}^{-1}\text{mol}^{-1}$

$\Delta G^\ominus = \Delta H^\ominus - T\Delta S^\ominus = 44 - 298 \times 0.119 = +8.5\,\text{kJ}\,\text{mol}^{-1}$

The reaction is endothermic, so the energy drive favours the reverse reaction. The entropy drive is positive, so it favours the forward direction.

9780170449564

Because Gibbs free energy is between -10 and $+10\,\text{kJ mol}^{-1}$, the reaction is an equilibrium reaction. It will go to some extent in both the forward and reverse directions but will not go to

~continued in right column ▲

completion under standard conditions. Therefore, the statement is partly correct.

2

Reaction	Energy drive direction	Entropy drive direction	Prediction
A	Forward	Forward	Reaction will proceed in the direction written. Both drives are in the forward direction.
B	Reverse	Forward	Unable to determine because the drives are in different directions and there is only a small entropy increase due to the product being more complex.
C	Reverse	Reverse	Reaction will proceed in the opposite direction to how it is written. Both drives are in the reverse direction.
D	Forward	Forward	Reaction will proceed in the direction written. Both drives are in the forward direction.
E	Reverse	Forward	Unable to determine because the drives are in different directions. Possibly the forward direction because the reactant has only gases while the product has a gas and a liquid.

3 a Reaction A:
$$\Delta G^{\ominus} = \Delta H^{\ominus} - T\Delta S^{\ominus} = -98 - (298 \times 63 \times 10^{-3})$$
$$= -116\,\text{kJ mol}^{-1}$$
Reaction B:
$$\Delta G^{\ominus} = \Delta H^{\ominus} - T\Delta S^{\ominus} = +180 - (298 \times 25 \times 10^{-3})$$
$$= +173\,\text{kJ mol}^{-1}$$
Reaction C:
$$\Delta G^{\ominus} = \Delta H^{\ominus} - T\Delta S^{\ominus} = +124 - (298 \times -209 \times 10^{-3})$$
$$= +130\,\text{kJ mol}^{-1}$$
Reaction D:
$$\Delta G^{\ominus} = \Delta H^{\ominus} - T\Delta S^{\ominus} = -80 - (298 \times 121 \times 10^{-3})$$
$$= -14\,\text{kJ mol}^{-1}$$
Reaction E:
$$\Delta G^{\ominus} = \Delta H^{\ominus} - T\Delta S^{\ominus} = +36 - (298 \times -199 \times 10^{-3})$$
$$= +95\,\text{kJ mol}^{-1}$$

b Reaction B was unable to be determined; however, $\Delta G^{\ominus} = +173\,\text{kJ mol}^{-1}$ so the reaction does not proceed in the direction given; it proceeds in the opposite direction.

Reaction E was thought to go in the forward direction; however, $\Delta G^{\ominus} = +95\,\text{kJ mol}^{-1}$ so the reaction does not proceed in the direction given; it proceeds in the opposite direction.

c Reactions A and D will spontaneously proceed in the directions as written while Reactions B, C and E will proceed spontaneously in the opposite direction to that which is written.

4 $\Delta H^{\ominus} = \Delta H^{\ominus}(\text{K}_2\text{CO}_3(\text{s})) - \{2 \times \Delta H^{\ominus}(\text{K}^+(\text{aq})) + \Delta H^{\ominus}(\text{CO}_3^{2-}(\text{aq}))\}$
$= -1151 - \{2 \times -252 + -675\} = +28\,\text{kJ mol}^{-1}$
$\Delta S^{\ominus} = S^{\ominus}(\text{K}_2\text{CO}_3(\text{s})) - \{2 \times S^{\ominus}(\text{K}^+(\text{aq})) + S^{\ominus}(\text{CO}_3^{2-}(\text{aq}))\}$
$= 156 - \{2 \times 101 + -50\} = +4\,\text{J K}^{-1}\,\text{mol}^{-1}$
$= 0.004\,\text{kJ K}^{-1}\,\text{mol}^{-1}$
$\Delta G^{\ominus} = \Delta H^{\ominus} - T\Delta S^{\ominus} = 28 - 298 \times 0.004 = +27\,\text{kJ mol}^{-1}$
The precipitate will not form. If $\text{K}_2\text{CO}_3(\text{s})$ is placed in water, it will spontaneously dissolve.
The student forgot to double the enthalpy and entropy values for $\text{K}^+(\text{aq})$.

5 a $\Delta H^{\ominus} = \Delta H^{\ominus}(\text{Sn(grey)}) - \Delta H^{\ominus}(\text{Sn(silver)}) = -2 - 0 = -2\,\text{kJ mol}^{-1}$
$\Delta S^{\ominus} = S^{\ominus}(\text{Sn(grey)}) - S^{\ominus}(\text{Sn(silver)}) = 44 - 52 = -8\,\text{J K}^{-1}\,\text{mol}^{-1}$
i At 22°C (295 K):
$$\Delta G^{\ominus} = \Delta H^{\ominus} - T\Delta S^{\ominus} = -2 - (295 \times -8 \times 10^{-3})$$
$$= 0.4\,\text{kJ mol}^{-1}$$

ii At -30°C (243 K):
$$\Delta G^{\ominus} = \Delta H^{\ominus} - T\Delta S^{\ominus} = -2 - (243 \times -8 \times 10^{-3})$$
$$= -0.06\,\text{kJ mol}^{-1}$$

b The allotrope change from silver to grey will proceed more as the temperature becomes lower.

c The values of $\Delta H^{\ominus}$ and $S^{\ominus}$ used for the calculation are at 298 K and assumed to be independent of temperature.

6 a $\Delta H^{\ominus} = \Delta H^{\ominus}(\text{CaO(s)}) + \Delta H^{\ominus}(\text{CO}_2(\text{g})) - \Delta H^{\ominus}(\text{CaCO}_3(\text{s}))$
$= -635 + (-394) - (-1207) = +178\,\text{kJ mol}^{-1}$
$\Delta S^{\ominus} = S^{\ominus}(\text{CaO(s)}) + S^{\ominus}(\text{CO}_2(\text{g})) - S^{\ominus}(\text{CaCO}_3(\text{s}))$
$= 38 + 214 - 93 = +159\,\text{J K}^{-1}\,\text{mol}^{-1} = 0.159\,\text{kJ K}^{-1}\,\text{mol}^{-1}$
$\Delta G^{\ominus} = \Delta H^{\ominus} - T\Delta S^{\ominus} = 178 - 298 \times 0.159 = +131\,\text{kJ mol}^{-1}$
The value of $\Delta G^{\ominus}$ is positive so the reaction will not proceed under normal conditions.

b Assume $\Delta G^{\ominus} = 0$
$\Delta G^{\ominus} = \Delta H^{\ominus} - T\Delta S^{\ominus}$
$0 = 178 - T \times 0.159$
$$T = \frac{178}{0.159} = 1119\,\text{K}$$

Therefore, calcium carbonate will decompose when the temperature is greater than 1119 K.

WS 15.3 PAGE 174

1 a This reaction is endothermic, so the energy drive is for the reverse direction. Since the reaction proceeds as written, the entropy drive must be in the forward direction and must be greater than the energy drive.

b **i** $\text{Mg}^{2+}(\text{aq}) + 2\text{OH}^-(\text{aq}) \rightarrow \text{Mg(OH)}_2(\text{s})$
ii $\Delta S^{\ominus} = S^{\ominus}(\text{Mg(OH)}_2(\text{s})) - \{S^{\ominus}(\text{Mg}^{2+}(\text{aq})) + 2 \times S^{\ominus}(\text{OH}^-(\text{aq}))\}$
$= 63 - \{-137 + 2 \times -11\} = +222\,\text{J K}^{-1}\,\text{mol}^{-1}$

c The spectator ions are not involved in the precipitation reaction so their entropy does not need to be included.

d The entropies of the precipitates are all similar; however, the solubilities are quite different. There does not appear to be a relationship between entropy of these precipitates and their solubility.

2 a Water is a polar molecule, so the positive end is attracted to anions and the negative end is attracted to the cations, forming ion–molecule bonds. This creates a higher degree of order within the solution as there are fewer water molecules able to move randomly.

When the precipitate forms, these water molecules are released and are able to move randomly within the solution, creating more disorder.

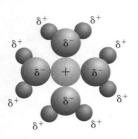 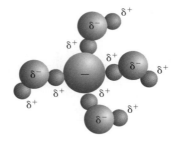

b A negative entropy means there is less disorder in the system. For the Mg^{2+} ion, this means the bonding of the ion to water has added order to the system, or the system has become less disordered.

3 For the reaction to occur spontaneously, ΔG for the reaction must be less than zero. As the reaction is endothermic, entropy (ΔS) must be positive and the term $T\Delta S$ must be greater than the positive ΔH value of 80.

4 a $\Delta H^{\ominus} = (2 \times -46 + 10 \times -286 + -895) - (-3345 + 2 \times -314)$
$\Delta H^{\ominus} = 126$ kJ
$\Delta S^{\ominus} = (2 \times 192 + 10 \times 70 + 124) - (427 + 2 \times 95)$
$\Delta S^{\ominus} = 591$ J K^{-1} = 0.591 kJ K^{-1}

Calculating the free energy from the above values at 298 K:
$\Delta G^{\ominus} = \Delta H^{\ominus} - T\Delta S^{\ominus} = 126 - 298 \times 0.591 = -50$ kJ

b $\Delta S^{\ominus}$ is positive, so decreasing temperature will cause the $T\Delta S^{\ominus}$ to become smaller and therefore $\Delta G^{\ominus}$ will increase. If the temperature is decreased enough, the reaction will cease to occur. Decreasing temperature also lowers the rate at which the reaction occurs.

MODULE FOUR: CHECKING UNDERSTANDING PAGE 177

1 A-6, B-16, C-10, D-2, E-7, F-11, G-13, H-8, I-4, J-1, K-14, L-9, M-15, N-5, O-12, P-3

2 a To the right, because one particle of a solid is being changed into 6 particles and a liquid is forming.

b To the left, because there are gas particles while on the right there is only a solid. Also, there are two particles on the left and only one on the right.

c To the right, because two particles are forming three particles.

d Cannot determine, because all particles are gases and there is the same number of particles on both sides.

3 a $2C_6H_{14}(l) + 19O_2(g) \rightarrow 12CO_2(g) + 14H_2O(g)$

b Heat absorbed by water
$= 700 \times 4.18 \times (36.3 - 14.7) = 6.32 \times 10^4$ J = 63.2 kJ
Heat absorbed by aluminium can
$= 58.3 \times 0.90 \times (36.3 - 14.7) = 1.13 \times 10^3$ J = 1.13 kJ
Total heat absorbed = 64.3 kJ
Mass of hexane used = 164.2 − 162.3 g = 1.9 g

Heat of combustion per g $= \dfrac{-64.3}{1.9} = -33.8$ kJ g^{-1}

n(hexane) $= \dfrac{1.9}{86} = 0.0221$ mol

Molar heat of combustion $= \dfrac{-64.3}{0.0221} = -2.91 \times 10^3$ kJ mol^{-1}

c Bonds to be broken:
- for each hexane molecule are $5 \times$ C−C, $14 \times$ C−H
- for each oxygen molecule $1 \times$ O=O.

Bonds to be formed:
- for each CO_2 molecule are $2 \times$ C=O
- for each H_2O molecule are $2 \times$ O−H.

From the balanced equation in part **a**, for 2 moles of hexane:
$\Delta H^{\ominus} = [2(5 \times B_{C-C} + 14 \times B_{C-H}) + 19 \times B_{O=O}]$
$\quad - [12(2 \times B_{C=O}) + 14(2 \times B_{O-H})]$
$\quad = [10 \times 348 + 28 \times 413 + 19 \times 498]$
$\quad - [24 \times 805 + 28 \times 463]$
$\quad = 24\,506 - 32\,284$
$\quad = -7778$ kJ

For 1 mole $\Delta H^{\ominus} = -3889$ kJ.

d The experimental value calculated in part **b** is likely to be the least accurate. Experimental errors in measurement of volumes, mass and temperature as well as heat loss to the environment during the experiment would have contributed to the lower value obtained.

The bond energy value is also less than the data table value. This is because the bond energies are averages over a wide range of compounds so are not specific to the bonds in particular compounds. Also, hexane is in liquid form so intermolecular bonds would need to be broken as well and these are not included in the calculation.

4 a Add the first two equations, deleting species that appear on both sides.
Reverse 3rd equation and add.

$\Delta H_1 = -37$ kJ $\qquad\qquad\qquad \Delta H_2 = -46$ kJ
$\cancel{N_2(g)} + \cancel{3H_2(g)} \rightarrow 2NH_3(g)$

$\Delta H_1 = -37$ kJ $\qquad\qquad\qquad \Delta H_2 = -46$ kJ
$\cancel{N_2(g)} + \cancel{3H_2(g)} \rightarrow 2NH_3(g)$

$N_2H_4(l) + H_2(g) \rightarrow 2NH_3(g) \qquad \Delta H_5 = -18$ kJ

b **i** $\Delta H = -18$ kJ
$\Delta S^{\ominus} = 2 \times S^{\ominus}(NH_3(g)) - \{S^{\ominus}(H_2(g)) + S^{\ominus}(N_2H_4(l))\}$
$\quad = 2 \times 193 - (130.7 + 121.2)$
$\quad = +134$ J K^{-1} mol^{-1}
$\quad = 0.134$ kJ K^{-1} mol^{-1}
$\Delta G^{\ominus} = \Delta H^{\ominus} - T\Delta S^{\ominus}$
$\quad = -18 - 298 \times 0.134$
$\quad = -57.9$ kJ mol^{-1}

ii Yes, the reaction is spontaneous at 25°C.

5 $\Delta G^{\ominus} = \Delta G^{\ominus}$(products) $- \Delta G^{\ominus}$(reaction)
$\quad = (-110 + -395) - (-33 + -409) = -63$ kJ mol^{-1}
$\Delta H^{\ominus} = \Delta H^{\ominus}$products $- \Delta H^{\ominus}$reactants
$\quad = (-127 + -495) - (-124 + -437) = -61$ kJ mol^{-1}
$\Delta S^{\ominus} = S^{\ominus}$products $- S^{\ominus}$reactants $= (S^{\ominus}AgCl + 133) - (141 + 83)$
$\quad = (S^{\ominus}AgCl - 91) \times 10^{-3}$ kJ K^{-1} mol^{-1}
$\Delta G^{\ominus} = \Delta H^{\ominus} - T\Delta S^{\ominus}$
$-63 = -61 - 298 \times ((S^{\ominus}AgCl - 91) \times 10^{-3})$
$-2 = -298((S^{\ominus}AgCl + 91) \times 10^{-3})$
$S^{\ominus}AgCl = 97.7$ J K^{-1} mol^{-1}

9780170449564